Teacher Edition

Eureka Math
Grade 8
Module 5

Special thanks go to the Gordan A. Cain Center and to the Department of Mathematics at Louisiana State University for their support in the development of *Eureka Math*.

For a free *Eureka Math* Teacher
Resource Pack, Parent Tip
Sheets, and more please
visit www.Eureka.tools

Published by the non-profit Great Minds

Copyright © 2015 Great Minds. All rights reserved. No part of this work may be reproduced or used in any form or by any means — graphic, electronic, or mechanical, including photocopying or information storage and retrieval systems — without written permission from the copyright holder. "Great Minds" and "Eureka Math" are registered trademarks of Great Minds.

Printed in the U.S.A.
This book may be purchased from the publisher at eureka-math.org
10 9 8 7 6 5 4 3

ISBN 978-1-63255-623-3

Eureka Math: A Story of Ratios **Contributors**

Michael Allwood, Curriculum Writer
Tiah Alphonso, Program Manager—Curriculum Production
Catriona Anderson, Program Manager—Implementation Support
Beau Bailey, Curriculum Writer
Scott Baldridge, Lead Mathematician and Lead Curriculum Writer
Bonnie Bergstresser, Math Auditor
Gail Burrill, Curriculum Writer
Beth Chance, Statistician
Joanne Choi, Curriculum Writer
Jill Diniz, Program Director
Lori Fanning, Curriculum Writer
Ellen Fort, Math Auditor
Kathy Fritz, Curriculum Writer
Glenn Gebhard, Curriculum Writer
Krysta Gibbs, Curriculum Writer
Winnie Gilbert, Lead Writer / Editor, Grade 8
Pam Goodner, Math Auditor
Debby Grawn, Curriculum Writer
Bonnie Hart, Curriculum Writer
Stefanie Hassan, Lead Writer / Editor, Grade 8
Sherri Hernandez, Math Auditor
Bob Hollister, Math Auditor
Patrick Hopfensperger, Curriculum Writer
Sunil Koswatta, Mathematician, Grade 8
Brian Kotz, Curriculum Writer
Henry Kranendonk, Lead Writer / Editor, Statistics
Connie Laughlin, Math Auditor
Jennifer Loftin, Program Manager—Professional Development
Nell McAnelly, Project Director
Ben McCarty, Mathematician
Stacie McClintock, Document Production Manager
Saki Milton, Curriculum Writer
Pia Mohsen, Curriculum Writer
Jerry Moreno, Statistician
Ann Netter, Lead Writer / Editor, Grades 6–7
Sarah Oyler, Document Coordinator
Roxy Peck, Statistician, Lead Writer / Editor, Statistics
Terrie Poehl, Math Auditor
Kristen Riedel, Math Audit Team Lead
Spencer Roby, Math Auditor
Kathleen Scholand, Math Auditor
Erika Silva, Lead Writer / Editor, Grade 6–7
Robyn Sorenson, Math Auditor
Hester Sutton, Advisor / Reviewer Grades 6–7
Shannon Vinson, Lead Writer / Editor, Statistics
Allison Witcraft, Math Auditor

Julie Wortmann, Lead Writer / Editor, Grade 7
David Wright, Mathematician, Lead Writer / Editor, Grades 6–7

Board of Trustees

Lynne Munson, President and Executive Director of Great Minds
Nell McAnelly, Chairman, Co-Director Emeritus of the Gordon A. Cain Center for STEM Literacy at Louisiana State University
William Kelly, Treasurer, Co-Founder and CEO at ReelDx
Jason Griffiths, Secretary, Director of Programs at the National Academy of Advanced Teacher Education
Pascal Forgione, Former Executive Director of the Center on K-12 Assessment and Performance Management at ETS
Lorraine Griffith, Title I Reading Specialist at West Buncombe Elementary School in Asheville, North Carolina
Bill Honig, President of the Consortium on Reading Excellence (CORE)
Richard Kessler, Executive Dean of Mannes College the New School for Music
Chi Kim, Former Superintendent, Ross School District
Karen LeFever, Executive Vice President and Chief Development Officer at ChanceLight Behavioral Health and Education
Maria Neira, Former Vice President, New York State United Teachers

This page intentionally left blank

Mathematics Curriculum

Table of Contents[1]

Examples of Functions from Geometry

Topics A–B (assessment 1 day, return 1 day, remediation or further applications 2 days)

---

[1]Each lesson is ONE day, and ONE day is considered a 45-minute period.

Module 5:   Examples of Functions from Geometry   **1**

This work is derived from Eureka Math ™ and licensed by Great Minds. ©2015 Great Minds. eureka-math.org
G8-M5-TE-B4-1.3.0-07.2015

Grade 8 • Module 5

Examples of Functions from Geometry

OVERVIEW

In Module 5, Topic A, students learn the concept of a function and why functions are necessary for describing geometric concepts and occurrences in everyday life. The module begins by explaining the important role functions play in making predictions. For example, if an object is dropped, a function allows us to determine its height at a specific time. To this point, student work has relied on assumptions of constant rates; here, students are given data that show that objects do not always travel at a constant speed. Once the concept of a function is explained, a formal definition of function is provided. A function is defined as an assignment to each input, exactly one output (**8.F.A.1**). Students learn that the assignment of some functions can be described by a mathematical rule or formula. With the concept and definition firmly in place, students begin to work with functions in real-world contexts. For example, students relate constant speed and other proportional relationships (**8.EE.B.5**) to linear functions. Next, students consider functions of discrete and continuous rates and understand the difference between the two. For example, students are asked to explain why they can write a cost function for a book, but they cannot input 2.6 into the function and get an accurate cost as the output.

Students apply their knowledge of linear equations and their graphs from Module 4 (**8.EE.B.5**, **8.EE.B.6**) to graphs of linear functions. Students know that the definition of a graph of a function is the set of ordered pairs consisting of an input and the corresponding output (**8.F.A.1**). Students relate a function to an input-output machine: a number or piece of data, known as the input, goes into the machine, and a number or piece of data, known as the output, comes out of the machine. In Module 4, students learned that a linear equation graphs as a line and that all lines are graphs of linear equations. In Module 5, students inspect the rate of change of linear functions and conclude that the rate of change is the slope of the graph of a line. They learn to interpret the equation $y = mx + b$ (**8.EE.B.6**) as defining a linear function whose graph is a line (**8.F.A.3**). Students also gain some experience with nonlinear functions, specifically by compiling and graphing a set of ordered pairs and then by identifying the graph as something other than a straight line.

Once students understand the graph of a function, they begin comparing two functions represented in different ways (**8.EE.C.8**), similar to comparing proportional relationships in Module 4. For example, students are presented with the graph of a function and a table of values that represent a function and are asked to determine which function has the greater rate of change (**8.F.A.2**). Students are also presented with functions in the form of an algebraic equation or written description. In each case, students examine the average rate of change and know that the one with the greater rate of change must overtake the other at some point.

In Topic B, students use their knowledge of volume from previous grade levels (**5.MD.C.3**, **5.MD.C.5**) to learn the volume formulas for cones, cylinders, and spheres (**8.G.C.9**). First, students are reminded of what they already know about volume, that volume is always a positive number that describes the hollowed-out portion of a solid figure that can be filled with water. Next, students use what they learned about the area of circles

This work is derived from Eureka Math ™ and licensed by Great Minds. ©2015 Great Minds. eureka-math.org
G8-M5-TE-B4-1.3.0-07.2015

(**7.G.B.4**) to determine the volume formulas for cones and cylinders. In each case, physical models are used to explain the formulas, beginning with a cylinder seen as a stack of circular disks that provide the height of the cylinder. Students consider the total area of the disks in three dimensions, understanding it as volume of a cylinder. Next, students make predictions about the volume of a cone that has the same dimensions as a cylinder. A demonstration shows students that the volume of a cone is one-third the volume of a cylinder with the same dimensions, a fact that will be proved in Module 7. Next, students compare the volume of a sphere to its circumscribing cylinder (i.e., the cylinder of dimensions that touches the sphere at points but does not cut off any part of it). Students learn that the formula for the volume of a sphere is two-thirds the volume of the cylinder that fits tightly around it. Students extend what they learned in Grade 7 (**7.G.B.6**) about how to solve real-world and mathematical problems related to volume from simple solids to include problems that require the formulas for cones, cylinders, and spheres.

Focus Standards

Define, evaluate, and compare functions.[2]

8.F.A.1   Understand that a function is a rule that assigns to each input exactly one output. The graph of a function is the set of ordered pairs consisting of an input and the corresponding output.[3]

8.F.A.2   Compare properties of two functions each represented in a different way (algebraically, graphically, numerically in tables, or by verbal descriptions). *For example, given a linear function represented by a table of values and a linear function represented by an algebraic expression, determine which function has the greater rate of change.*

8.F.A.3   Interpret the equation $y = mx + b$ as defining a linear function, whose graph is a straight line; give examples of functions that are not linear. *For example, the function $A = s^2$ giving the area of a square as a function of its side length is not linear because its graph contains the points $(1, 1)$, $(2, 4)$ and $(3, 9)$ which are not on a straight line.*

Solve real-world and mathematical problems involving volume of cylinders, cones, and spheres.

8.G.C.9[4]   Know the formulas for the volumes of cones, cylinders, and spheres and use them to solve real-world and mathematical problems.

---

[2]Linear and nonlinear functions are compared in this module using linear equations and area/volume formulas as examples.
[3]Function notation is not required in Grade 8.
[4]Solutions that introduce irrational numbers are not introduced until Module 7.

Module 5:   Examples of Functions from Geometry

This work is derived from Eureka Math ™ and licensed by Great Minds. ©2015 Great Minds. eureka-math.org
G8-M5-TE-B4-1.3.0-07.2015

Foundational Standards

Geometric measurement: Understand concepts of volume and relate volume to multiplication and to addition.

5.MD.C.3   Recognize volume as an attribute of solid figures and understand concepts of volume measurement.

     a.   A cube with side length 1 unit, called a "unit cube," is said to have "one cubic unit" of volume, and can be used to measure volume.

     b.   A solid figure which can be packed without gaps or overlaps using n unit cubes is said to have a volume of n cubic units.

5.MD.C.5   Relate volume to the operations of multiplication and addition and solve real world and mathematical problems involving volume.

     a.   Find the volume of a right rectangular prism with whole-number side lengths by packing it with unit cubes, and show that the volume is the same as would be found by multiplying the edge lengths, equivalently by multiplying the height by the area of the base. Represent threefold whole-number products as volumes, e.g., to represent the associative property of multiplication.

     b.   Apply the formulas $V = l \times w \times h$ and $V = b \times h$ for rectangular prisms to find volumes of right rectangular prisms with whole–number edge lengths in the context of solving real world and mathematical problems.

     c.   Recognize volume as additive. Find volume of solid figures composed of two non-overlapping right rectangular prisms by adding the volumes of the non-overlapping parts, applying this technique to real world problems.

Solve real-life and mathematical problems involving angle measure, area, surface area, and volume.

7.G.B.4   Know the formulas for the area and circumference of a circle and use them to solve problems; give an informal derivation of the relationship between the circumference and area of a circle.

7.G.B.6   Solve real-world and mathematical problems involving area, volume and surface area of two-and three-dimensional objects composed of triangles, quadrilaterals, polygons, cubes, and right prisms.

Understand the connections between proportional relationships, lines, and linear equations.

8.EE.B.5   Graph proportional relationships, interpreting the unit rate as the slope of the graph. Compare two different proportional relationships represented in different ways. *For example, compare a distance-time graph to a distance-time equation to determine which of two moving objects has greater speed.*

     Module 5:   Examples of Functions from Geometry

This work is derived from Eureka Math ™ and licensed by Great Minds. ©2015 Great Minds. eureka-math.org
G8-M5-TE-B4-1.3.0-07.2015

8.EE.B.6   Use similar triangles to explain why the slope m is the same between any two distinct points on a non-vertical line in the coordinate plane; derive the equation $y = mx$ for a line through the origin and the equation $y = mx + b$ for a line intercepting the vertical axis at b.

Analyze and solve linear equations and pairs of simultaneous linear equations.

8.EE.C.7   Solve linear equations in one variable.

     a.   Give examples of linear equations in one variable with one solution, infinitely many solutions, or no solutions. Show which of these possibilities is the case by successively transforming the given equation into simpler forms, until an equivalent equation of the form $x = a$, $a = a$, or $a = b$ results (where a and b are different numbers).

     b.   Solve linear equations with rational number coefficients, including equations whose solutions require expanding expressions using the distributive property and collecting like terms.

8.EE.C.8   Analyze and solve pairs of simultaneous linear equations.

     a.   Understand that solutions to a system of two linear equations in two variables correspond to points of intersection of their graphs, because points of intersection satisfy both equations simultaneously.

     b.   Solve systems of two linear equations in two variables algebraically, and estimate solutions by graphing the equations. Solve simple cases by inspection. *For example, $3x + 2y = 5$ and $3x + 2y = 6$ have no solution because $3x + 2y$ cannot simultaneously be* 5 *and* 6.

     c.   Solve real-world and mathematical problems leading to two linear equations in two variables. *For example, given coordinates for two pairs of points, determine whether the line through the first pair of points intersects the line through the second pair.*

Focus Standards for Mathematical Practice

MP.2   **Reason abstractly or quantitatively**. Students examine, interpret, and represent functions symbolically. They make sense of quantities and their relationships in problem situations. For example, students make sense of values as they relate to the total cost of items purchased or a phone bill based on usage in a particular time interval. Students use what they know about rate of change to distinguish between linear and nonlinear functions. Further, students contextualize information gained from the comparison of two functions.

MP.6   **Attend to precision**. Students use notation related to functions, in general, as well as notation related to volume formulas. Students are expected to clearly state the meaning of the symbols used in order to communicate effectively and precisely to others. Students attend to precision when they interpret data generated by functions. They know when claims are false; for example, calculating the height of an object after it falls for -2 seconds. Students also understand that a table of values is an incomplete *representation* of a continuous function, as an infinite number of values can be found for a function.

This work is derived from Eureka Math ™ and licensed by Great Minds. ©2015 Great Minds. eureka-math.org
G8-M5-TE-B4-1.3.0-07.2015

MP.8   **Look for and express regularity in repeated reasoning.** Students use repeated computations to determine equations from graphs or tables. While focused on the details of a specific pair of numbers related to the input and output of a function, students maintain oversight of the process. As students develop equations from graphs or tables, they evaluate the reasonableness of their equation as they ensure that the desired output is a function of the given input.

Terminology

New or Recently Introduced Terms

- **Cone**[8.G.9] (Let B be a polygonal region or a disk in a plane E, and V be a point not in E. The *cone with base B and vertex V* is the union of all segments $\overline{VP}$ for all points P in B.

  A cone is named by its base. If the base is a polygonal region, then the cone is usually called a *pyramid*. For example, a cone with a triangular region for its base is called a *triangular pyramid*. A cone with a circular region for its base whose vertex lies on the perpendicular line to the base that passes through the center of the circle is called a *right circular cone.* This cone is the one usually shown in elementary and middle school textbooks. The name of it is often shortened to just *cone*.)

- **Cylinder** (Let E and E' be two parallel planes, let B be a polygonal region or a disk in the plane E, and let L be a line which intersects E and E' but not B. At each point P of B, consider the segment $\overline{PP'}$ parallel to line L, joining P to a point P' of the plane E'. The union of all these segments is called a *cylinder with base B.* The regions B and B' are called the *base faces* (or just *bases*) of the prism.

  A cylinder is named by its base. If the base is a polygonal region, then the cylinder is usually called a *prism*. For example, a cylinder with a triangular region for its base is called a *triangular prism*. A cylinder with a circular region for its base that is defined by a line that is perpendicular to the base is called a *right circular cylinder.* This cylinder is the one usually shown in elementary and middle school textbooks, where the name is often shortened to just *cylinder*.)

  Equation Form of a Linear Function (description) (The *equation form of a linear function* is an equation of the form $y = mx + b$, where the number m is called the *rate of change of the linear function* and the number b is called the *initial value of the linear function.* To calculate the output named by the dependent variable y, an input is substituted into the independent variable x and evaluated.)

- **Function (description)** (A *function* is a correspondence between a set (whose elements are called *inputs*) and another set (whose elements are called *outputs*) such that each input corresponds to one and only one output. The correspondence is often given as a *rule*: the output is a number found by substituting an input number into the variable of a one-variable expression and evaluating.

  For example, a proportional relationship is a special type of function whose output is always given by multiplying the input number by another number (the constant of proportionality).)

- **Graph of a Linear Function** (The graph of a linear function represented by the equation $y = mx + b$ is the set of ordered pairs (x, y) for inputs x and outputs y that make the equation true. When the graph of a linear function is a line (i.e., the set of inputs is all real numbers), then m is the slope of the line and b is the y-intercept of the line.)

      Module 5:   Examples of Functions from Geometry

This work is derived from Eureka Math ™ and licensed by Great Minds. ©2015 Great Minds. eureka-math.org
G8-M5-TE-B4-1.3.0-07.2015

- **Lateral Edge and Face of a Prism** (Suppose the base B of a prism is a polygonal region and P_i is a vertex of B. Let P_i' be the corresponding point in B' such that $\overline{P_i P_i'}$ is parallel to the line L defining the prism. The segment $\overline{P_i P_i'}$ is called a *lateral edge* of the prism. If $\overline{P_i P_{i+1}}$ is a base edge of the base B (a side of B), and F is the union of all segments $\overline{PP'}$ parallel to line L for which points P are in segment $\overline{P_i P_{i+1}}$ and points P' are in B', then F is a *lateral face* of the prism. It can be shown that a lateral face of a prism is always a region enclosed by a parallelogram.)

- **Lateral Edge and Face of a Pyramid**[8.G.9] (Suppose the base B of a pyramid with vertex V is a polygonal region and P_i is a vertex of B. The segment $\overline{P_i V}$ is called a *lateral edge* of the pyramid. If $\overline{P_i P_{i+1}}$ is a base edge of the base B (a side of B), and F is the union of all segments $\overline{PV}$ for all points P in the segment $\overline{P_i P_{i+1}}$, then F is called a *lateral face* of the pyramid. It can be shown that the face of a pyramid is always a triangular region.)

- **Linear Function (description)** (A *linear function* is a function whose inputs and outputs are real numbers such that each output is given by substituting an input into a linear expression and evaluating.

- **Solid Sphere or Ball** (Given a point C in the 3-dimensional space and a number $r > 0$, the *solid sphere (or ball) with center C and radius r* is the set of all points in space whose distance from the point C is less than or equal to r.)

- **Sphere** (Given a point C in the 3-dimensional space and a number $r > 0$, the *sphere with center C and radius r* is the set of all points in space that are distance r from the point C.)

Familiar Terms and Symbols[5]

- Area
- Linear equation
- Nonlinear equation
- Rate of change
- Solids
- Volume

Suggested Tools and Representations

- 3D solids: cones, cylinders, and spheres

---

[5]These are terms and symbols students have seen previously.

Module 5:   Examples of Functions from Geometry   7

This work is derived from Eureka Math ™ and licensed by Great Minds. ©2015 Great Minds. eureka-math.org
G8-M5-TE-B4-1.3.0-07.2015

Rapid White Board Exchanges

Implementing a RWBE requires that each student be provided with a personal white board, a white board marker, and a means of erasing his work. An economic choice for these materials is to place sheets of card stock inside sheet protectors to use as the personal white boards and to cut sheets of felt into small squares to use as erasers.

A RWBE consists of a sequence of 10 to 20 problems on a specific topic or skill that starts out with a relatively simple problem and progressively gets more difficult. The teacher should prepare the problems in a way that allows her to reveal them to the class one at a time. A flip chart or PowerPoint presentation can be used, or the teacher can write the problems on the board and either cover some with paper or simply write only one problem on the board at a time.

The teacher reveals, and possibly reads aloud, the first problem in the list and announces, "Go." Students work the problem on their personal white boards as quickly as possible and hold their work up for their teacher to see their answers as soon as they have the answer ready. The teacher gives immediate feedback to each student, pointing and/or making eye contact with the student and responding with an affirmation for correct work such as, "Good job!", "Yes!", or "Correct!", or responding with guidance for incorrect work such as, "Look again," "Try again," "Check your work." In the case of the RWBE, it is not recommended that the feedback include the name of the student receiving the feedback.

If many students have struggled to get the answer correct, go through the solution of that problem as a class before moving on to the next problem in the sequence. Fluency in the skill has been established when the class is able to go through each problem in quick succession without pausing to go through the solution of each problem individually. If only one or two students have not been able to successfully complete a problem, it is appropriate to move the class forward to the next problem without further delay; in this case find a time to provide remediation to that student before the next fluency exercise on this skill is given.

Assessment Summary

Assessment Type	Administered	Format	Standards Addressed
End-of-Module Assessment Task	After Topic B	Constructed response with rubric	8.F.A.1, 8.F.A.2, 8.F.A.3, 8.G.C.9

This work is derived from Eureka Math ™ and licensed by Great Minds. ©2015 Great Minds. eureka-math.org
G8-M5-TE-B4-1.3.0-07.2015

Mathematics Curriculum

8 GRADE

Topic A

Functions

8.F.A.1, 8.F.A.2, 8.F.A.3

Focus Standards:	8.F.A.1	Understand that a function is a rule that assigns to each input exactly one output. The graph of a function is the set of ordered pairs consisting of an input and the corresponding output.
	8.F.A.2	Compare properties of two functions each represented in a different way (algebraically, graphically, numerically in tables, or by verbal descriptions). *For example, given a linear function represented by a table of values and a linear function represented by an algebraic expression, determine which function has the greater rate of change.*
	8.F.A.3	Interpret the equation $y = mx + b$ as defining a linear function, whose graph is a straight line; give examples of functions that are not linear. *For example, the function $A = s^2$ giving the area of a square as a function of its side length is not linear because its graph contains the points $(1, 1)$, $(2, 4)$ and $(3, 9)$, which are not on a straight line.*
Instructional Days:	8	

Lesson 1: The Concept of a Function (P)[1]

Lesson 2: Formal Definition of a Function (S)

Lesson 3: Linear Functions and Proportionality (P)

Lesson 4: More Examples of Functions (P)

Lesson 5: Graphs of Functions and Equations (E)

Lesson 6: Graphs of Linear Functions and Rate of Change (S)

Lesson 7: Comparing Linear Functions and Graphs (E)

Lesson 8: Graphs of Simple Nonlinear Functions (E)

---

[1]Lesson Structure Key: **P**-Problem Set Lesson, **M**-Modeling Cycle Lesson, **E**-Exploration Lesson, **S**-Socratic Lesson

Topic A: Functions

9

This work is derived from Eureka Math ™ and licensed by Great Minds. ©2015 Great Minds. eureka-math.org
G8-M5-TE-B4-1.3.0-07.2015

Lesson 1 relies on students' understanding of constant rate, a skill developed in previous grade levels and reviewed in Module 4 (**6.RP.A.3b**, **7.RP.A.2**). Students are confronted with the fact that the concept of constant rate, which requires the assumption that a moving object travels at a constant speed, cannot be applied to all moving objects. Students examine a graph and a table that demonstrate the nonlinear effect of gravity on a falling object. This example provides the reasoning for the need of functions. In Lesson 2, students continue their investigation of time and distance data for a falling object and learn that the scenario can be expressed by a formula. Students are introduced to the terms *input* and *output* and learn that a function assigns to each input exactly one output. Though students do not learn the traditional "vertical-line test," students know that the graph of a function is the set of ordered pairs consisting of an input and the corresponding output. Students also learn that not all functions can be expressed by a formula, but when they are, the function rule allows us to make predictions about the world around us. For example, with respect to the falling object, the function allows us to predict the height of the object for any given time interval.

In Lesson 3, constant rate is revisited as it applies to the concept of linear functions and proportionality in general. Lesson 4 introduces students to the fact that not all rates are continuous. That is, a cost function for the cost of a book can be written, yet the cost of 3.6 books cannot realistically be found. Students are also introduced to functions that do not use numbers at all, as in a function where the input is a card from a standard deck, and the output is the suit.

Lesson 5 is when students begin graphing functions of two variables. Students graph linear and nonlinear functions, and the guiding question of the lesson, "Why not just look at graphs of equations in two variables?" is answered because not all graphs of equations are graphs of functions. Students continue their work on graphs of linear functions in Lesson 6. In this lesson, students investigate the rate of change of functions and conclude that the rate of change for linear functions is the slope of the graph. In other words, this lesson solidifies the fact that the equation $y = mx + b$ defines a linear function whose graph is a straight line.

With the knowledge that the graph of a linear function is a straight line, students begin to compare properties of two functions that are expressed in different ways in Lesson 7. One example of this relates to a comparison of phone plans. Students are provided a graph of a function for one plan and an equation of a function that represents another plan. In other situations, students are presented with functions that are expressed algebraically, graphically, and numerically in tables, or are described verbally. Students must use the information provided to answer questions about the rate of change of each function. In Lesson 8, students work with simple nonlinear functions of area and volume and their graphs.

EUREKA MATH

This work is derived from Eureka Math ™ and licensed by Great Minds. ©2015 Great Minds. eureka-math.org
G8-M5-TE-B4-1.3.0-07.2015

Lesson 1: The Concept of a Function

Student Outcomes

- Students analyze a nonlinear data set.
- Students realize that an assumption of a constant rate of motion is not appropriate for all situations.

Lesson Notes

Using up-to-date data can add new elements of life to a lesson for students. If time permits, consider gathering examples of data from the Internet, and use that data as examples throughout this topic.

Much of the discussion in this module is based on parts from the following sources:

H. Wu, *Introduction to School Algebra*, http://math.berkeley.edu/~wu/Algebrasummary.pdf

H. Wu, *Teaching Geometry in Grade 8 and High School According to the Common Core Standards*,

https://math.berkeley.edu/~wu/CCSS-Geometry_1.pdf

Classwork

Discussion (4 minutes)

- In the last module we focused on situations that were worked with two varying quantities, one changing with respect to the other according to some constant rate of change factor. Consequently, each situation could be analyzed with a linear equation. Such a formula then gave us the means to determine the value of the one quantity given a specific value of the other.

- There are many situations, however, for which assuming a constant rate of change relationship between two quantities is not appropriate. Consequently, there is no linear equation to describe their relationship. If we are fortunate, we may be able to find a mathematical equation describing their relationship nonetheless.

- Even if this is not possible, we may be able to use data from the situation to see a relationship of some kind between the values of one quantity and their matching values of the second quantity.

- Mathematicians call any clearly-described rule that assigns to each value of one quantity a single value of a second quantity a *function*. Functions could be described by words ("Assign to whole number its first digit," for example), by a table of values or by a graph (a graph or table of your height at different times of your life, for example) or, if we are lucky, by a formula. (For instance, the formula $y = 2x$ describes the rule "Assign to each value x double its value.")

- All becomes clear as this topic progresses. We start by looking at one curious set of data.

Lesson 1:   The Concept of a Function   11

This work is derived from Eureka Math ™ and licensed by Great Minds. ©2015 Great Minds. eureka-math.org
G8-M5-TE-B4-1.3.0-07.2015

Example 1 (7 minutes)

This example is used to point out that, contrary to much of the previous work in assuming a constant rate relationship exists, students cannot, in general, assume this is always the case. Encourage students to make sense of the problem and attempt to solve it on their own. The goal is for students to develop an understanding of the subtleties predicting data values.

Example 1

Suppose a moving object travels 256 feet in 4 seconds. Assume that the object travels at a constant speed, that is, the motion of the object can be described by a linear equation. Write a linear equation in two variables to represent the situation, and use the equation to predict how far the object has moved at the four times shown.

Number of seconds in motion (x)	Distance traveled in feet (y)
1	64
2	128
3	192
4	256

MP.1

- Suppose a moving object travels 256 feet in 4 seconds. Assume that the object travels at a constant speed, that is, the motion of the object can be described by a linear equation. Write a linear equation in two variables to represent the situation, and use the equation to predict how far the object has moved at the four times shown.

    - *Let x represent the time it takes to travel y feet.*

$$\frac{256}{4} = \frac{y}{x}$$

$$y = \frac{256}{4} x$$

$$y = 64x$$

- What are some of the predictions that this equation allows us to make?

    - *After one second, or when $x = 1$, the distance traveled is 64 feet.*

Accept any reasonable predictions that students make.

- Use your equation to complete the table.
- What is the average speed of the moving object from 0 to 3 seconds?

    - *The average speed is 64 feet per second. We know that the object has a constant rate of change; therefore, we expect the average speed to be the same over any time interval.*

Example 2 (15 minutes)

- Suppose I now reveal that the object is a stone being dropped from a height of 256 feet. It takes exactly 4 seconds for the stone to hit the ground. Do you think we can assume constant speed in this situation? Is our linear equation describing the situation still valid?

    Lesson 1:   The Concept of a Function

EUREKA MATH™

This work is derived from Eureka Math ™ and licensed by Great Minds. ©2015 Great Minds. eureka-math.org
G8-M5-TE-B4-1.3.0-07.2015

Example 2

The object, a stone, is dropped from a height of 256 feet. It takes exactly 4 seconds for the stone to hit the ground. How far does the stone drop in the first 3 seconds? What about the last 3 seconds? Can we assume constant speed in this situation? That is, can this situation be expressed using a linear equation?

Number of seconds (x)	Distance traveled in feet (y)
1	16
2	64
3	144
4	256

Provide students time to discuss this in pairs. Lead a discussion in which students share their thoughts with the class. It is likely they will say the motion of a falling object is linear and that the work conducted in the previous example is appropriate.

- If this is a linear situation, then we predict that the stone drops 192 feet in the first 3 seconds.

Now consider viewing the 10-second "ball drop" video at the following link: http://www.youtube.com/watch?v=KrX_zLuwOvc. Consider showing it more than once.

- If we were to slow the video down and record the distance the ball dropped after each second, here is the data we would obtain:

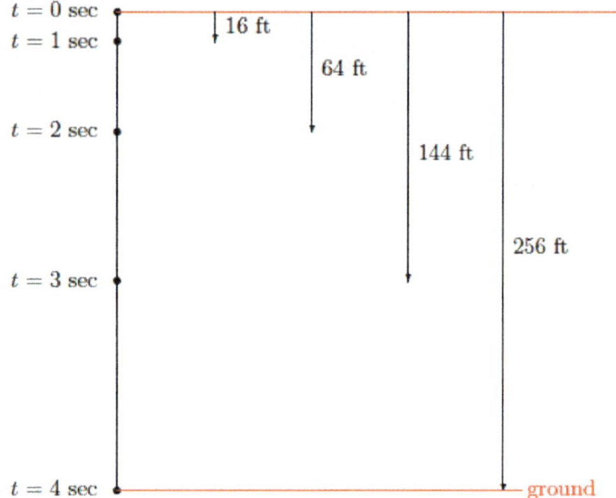

Have students record the data in the table of Example 2.

- Was the linear equation developed in Example 1 appropriate after all?

Students who thought the stone was traveling at constant speed should realize that the predictions were not accurate for this situation. Guide their thinking using the discussion points below.

- According to the data, how many feet did the stone drop in 3 seconds?
  - *The stone dropped* 144 *feet.*
- How can that be? It must be that our initial assumption of constant rate was incorrect.
  What predictions can we make now?
  - *After one second, $x = 1$; the stone dropped* 16 *feet, etc.*

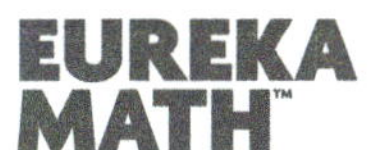

This work is derived from Eureka Math ™ and licensed by Great Minds. ©2015 Great Minds. eureka-math.org
G8-M5-TE-B4-1.3.0-07.2015

- Let's make a prediction based on a value of x that is not listed in the table. How far did the stone drop in the first 3.5 seconds? What have we done in the past to figure something like this out?

  - *We wrote a proportion using the known times and distances.*

Allow students time to work with proportions. Encourage them to use more than one pair of data values to determine an answer. Some students might suggest they cannot use proportions for this work as they have just ascertained that there is not a constant rate of change. Acknowledge this. The work with proportions some students do will indeed confirm this.

  - *Sample student work:*

    Let x be the distance, in feet, the stone drops in 3.5 seconds.

$$\frac{16}{1} = \frac{x}{3.5} \qquad\qquad \frac{64}{2} = \frac{x}{3.5} \qquad\qquad \frac{144}{3} = \frac{x}{3.5}$$

$$x = 56 \qquad\qquad\qquad 2x = 224 \qquad\qquad\qquad 3x = 504$$

$$x = 112 \qquad\qquad\qquad x = 168$$

MP.3

- Is it reasonable that the stone would drop 56 feet in 3.5 seconds? Explain.

  - *No, it is not reasonable. Our data shows that after 2 seconds, the stone has already dropped 64 feet. Therefore, it is impossible that it could have only dropped 56 feet in 3.5 seconds.*

- What about 112 feet in 3.5 seconds? How reasonable is that answer? Explain.

  - *The answer of 112 feet in 3.5 seconds is not reasonable either. The data shows that the stone dropped 144 feet in 3 seconds.*

- What about 168 feet in 3.5 seconds? What do you think about that answer? Explain.

  - *That answer is the most likely because at least it is greater than the recorded 144 feet in 3 seconds.*

- What makes you think that the work done with a third proportion will give us a correct answer when the first two did not? Can we rely on this method for determining an answer?

  - *This does not seem to be a reliable method. If we had only done one computation and not evaluated the reasonableness of our answer, we would have been wrong.*

- What this means is that the table we used does not tell the whole story about the falling stone. Suppose, by repeating the experiment and gathering more data of the motion, we obtained the following table:

Number of seconds (x)	Distance traveled in feet (y)
0.5	4
1	16
1.5	36
2	64
2.5	100
3	144
3.5	196
4	256

- Were any of the predictions we made about the distance dropped during the first 3.5 seconds correct?

     Lesson 1:   The Concept of a Function

EUREKA MATH™

This work is derived from Eureka Math ™ and licensed by Great Minds. ©2015 Great Minds. eureka-math.org
G8-M5-TE-1.3.0-07.2015

Have a discussion with students about why we want to make predictions at all. Students should recognize that making predictions helps us make sense of the world around us. Some scientific discoveries began with a prediction, then an experiment to prove or disprove the prediction, and then were followed by some conclusion.

- Now it is clear that none of our answers for the distance traveled in 3.5 seconds were correct. In fact, the stone dropped 196 feet in the first 3.5 seconds. Does the table on the previous page capture the motion of the stone completely? Explain?

  - *No. There are intervals of time between those in the table. For example, the distance it drops in 1.6 seconds is not represented.*

- If we were to record the data for every 0.1 second that passed, would that be enough to capture the motion of the stone?

  - *No. There would still be intervals of time not represented. For example, 1.61 seconds.*

- To tell the whole story, we would need information about where the stone is after the first t seconds for *every* t satisfying $0 \le t \le 4$.

- Nonetheless, for each value t for $0 \le t \le 4$, there is some specific value to assign to that value: the distance the stone fell during the first t seconds. That it, there is some rule that assigns to each value t between $0 \le t \le 4$ a real number. We have an example of a *function*. (And, at present, we don't have a mathematical formula describing that function.)

Some students, however, might observe at this point the data does seem to be following the formula $y = 16x^2$.

Exercises 1–6 (10 minutes)

Students complete Exercises 1–6 in pairs or small groups.

Exercises 1–6

Use the table to answer Exercises 1–5.

Number of seconds (x)	Distance traveled in feet (y)
0.5	4
1	16
1.5	36
2	64
2.5	100
3	144
3.5	196
4	256

1. Name two predictions you can make from this table.

   Sample student responses:

   After 2 seconds, the object traveled 64 feet. After 3.5 seconds, the object traveled 196 feet.

This work is derived from Eureka Math ™ and licensed by Great Minds. ©2015 Great Minds. eureka-math.org
G8-M5-TE-B4-1.3.0-07.2015

2. Name a prediction that would require more information.

   Sample student response:

   We would need more information to predict the distance traveled after 3.75 seconds.

3. What is the average speed of the object between 0 and 3 seconds? How does this compare to the average speed calculated over the same interval in Example 1?

   MP.2

$$\text{Average Speed} = \frac{\text{distance traveled over a given time interval}}{\text{time interval}}$$

   The average speed is 48 feet per second: $\frac{144}{3} = 48$. *This is different from the average speed calculated in Example 1. In Example 1, the average speed over an interval of 3 seconds was 64 feet per second.*

4. Take a closer look at the data for the falling stone by answering the questions below.

   a. How many feet did the stone drop between 0 and 1 second?

      The stone dropped 16 feet between 0 and 1 second.

   b. How many feet did the stone drop between 1 and 2 seconds?

      The stone dropped 48 feet between 1 and 2 seconds.

   c. How many feet did the stone drop between 2 and 3 seconds?

      The stone dropped 80 feet between 2 and 3 seconds.

   d. How many feet did the stone drop between 3 and 4 seconds?

      The stone dropped 112 feet between 3 and 4 seconds.

   e. Compare the distances the stone dropped from one time interval to the next. What do you notice?

      Over each interval, the difference in the distance was 32 feet. For example, $16 + 32 = 48$, $48 + 32 = 80$, and $80 + 32 = 112$.

5. What is the average speed of the stone in each interval 0.5 second? For example, the average speed over the interval from 3.5 seconds to 4 seconds is

$$\frac{\text{distance traveled over a given time interval}}{\text{time interval}} = \frac{256 - 196}{4 - 3.5} = \frac{60}{0.5} = 120; 120 \text{ feet per second}$$

   Repeat this process for every half-second interval. Then, answer the question that follows.

   a. Interval between 0 and 0.5 second:

      $\frac{4}{0.5} = 8; 8$ *feet per second*

   b. Interval between 0.5 and 1 second:

      $\frac{12}{0.5} = 24; 24$ *feet per second*

   c. Interval between 1 and 1.5 seconds:

      $\frac{20}{0.5} = 40; 40$ *feet per second*

   d. Interval between 1.5 and 2 seconds:

      $\frac{28}{0.5} = 56; 56$ *feet per second*

    Lesson 1:   The Concept of a Function

EUREKA MATH

This work is derived from Eureka Math ™ and licensed by Great Minds. ©2015 Great Minds. eureka-math.org
G8-M5-TE-B4-1.3.0-07.2015

e. Interval between 2 and 2.5 seconds:

$$\frac{36}{0.5} = 72; 72 \text{ feet per second}$$

f. Interval between 2.5 and 3 seconds:

$$\frac{44}{0.5} = 88; 88 \text{ feet per second}$$

g. Interval between 3 and 3.5 seconds:

$$\frac{52}{0.5} = 104; 104 \text{ feet per second}$$

h. Compare the average speed between each time interval. What do you notice?

Over each interval, there is an increase in the average speed of 16 feet per second. For example,

$8 + 16 = 24, \ 24 + 16 = 40, \ 40 + 16 = 56,$ *and so on.*

6. Is there any pattern to the data of the falling stone? Record your thoughts below.

Time of interval in seconds (t)	1	2	3	4
Distance stone fell in feet (y)	16	64	144	256

Accept any reasonable patterns that students notice as long as they can justify their claim. In the next lesson, students learn that $y = 16t^2$.

Each distance has 16 as a factor. For example, $16 = 1(16), \ 64 = 4(16), \ 144 = 9(16),$ and $256 = 16(16)$.

Closing (4 minutes)

Summarize, or ask students to summarize, the main points from the lesson:

- We know that we cannot always assume motion is given at a constant rate.

Lesson Summary

A *function* is a rule that assigns to each value of one quantity a single value of a second quantity Even though we might not have a formula for that rule, we see that functions do arise in real-life situations.

Exit Ticket (5 minutes)

This work is derived from Eureka Math ™ and licensed by Great Minds. ©2015 Great Minds. eureka-math.org
G8-M5-TE-B4-1.3.0-07.2015

Name ____   Date _______________________

Lesson 1: The Concept of a Function

Exit Ticket

A ball is bouncing across the school yard. It hits the ground at $(0,0)$ and bounces up and lands at $(1,0)$ and bounces again. The graph shows only one bounce.

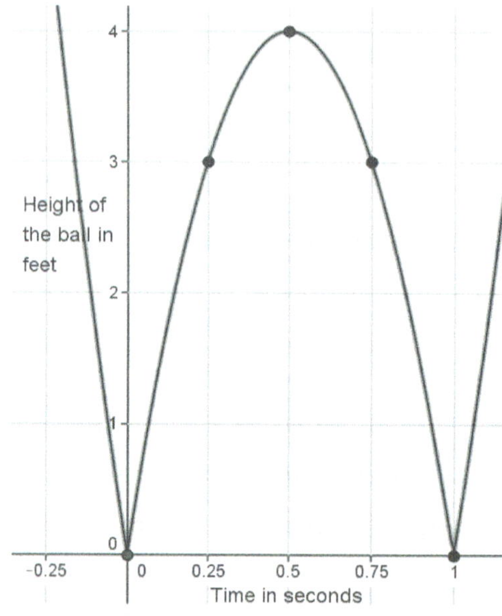

a.   Identify the height of the ball at the following values of t: 0, 0.25, 0.5, 0.75, 1.

b.   What is the average speed of the ball over the first 0.25 seconds? What is the average speed of the ball over the next 0.25 seconds (from 0.25 to 0.5 seconds)?

c.   Is the height of the ball changing at a constant rate?

This work is derived from Eureka Math ™ and licensed by Great Minds. ©2015 Great Minds. eureka-math.org
G8-M5-TE-B4-1.3.0-07.2015

EUREKA
MATH™

Exit Ticket Sample Solutions

A ball is bouncing across the school yard. It hits the ground at $(0, 0)$ and bounces up and lands at $(1, 0)$ and bounces again. The graph shows only one bounce.

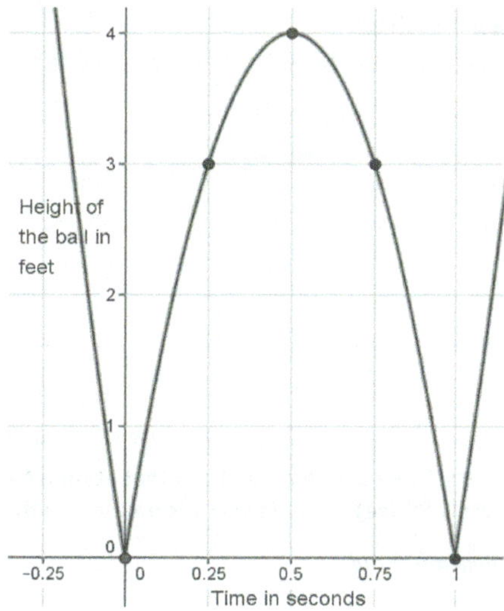

a.   **Identify the height of the ball at the following time values: $0, 0.25, 0.5, 0.75, 1$.**

When $t = 0$, the height of the ball is 0 feet above the ground. It has just hit the ground.

When $t = 0.25$, the height of the ball is 3 feet above the ground.

When $t = 0.5$, the height of the ball is 4 feet above the ground.

When $t = 0.75$, the height of the ball is 3 feet above the ground.

When $t = 1$, the height of the ball is 0 feet above the ground. It has hit the ground again.

b.   **What is the average speed of the ball over the first 0.25 seconds? What is the average speed of the ball over the next 0.25 seconds (from 0.25 to 0.5 seconds)?**

$$\frac{\text{distance traveled over a given time interval}}{\text{time interval}} = \frac{3-0}{0.25-0} = \frac{3}{0.25} = 12; 12 \text{ feet per second}$$

$$\frac{\text{distance traveled over a given time interval}}{\text{time interval}} = \frac{4-3}{0.5-.25} = \frac{1}{0.25} = 4; 4 \text{ feet per second}$$

c.   **Is the height of the ball changing at a constant rate?**

No, it is not. If the ball were traveling at a constant rate, the average speed would be the same over any time interval.

This work is derived from Eureka Math ™ and licensed by Great Minds. ©2015 Great Minds. eureka-math.org
G8-M5-TE-B4-1.3.0-07.2015

Problem Set Sample Solutions

A ball is thrown across the field from point A to point B. It hits the ground at point B. The path of the ball is shown in the diagram below. The x-axis shows the horizontal distance the ball travels in feet, and the y-axis shows the height of the ball in feet. Use the diagram to complete parts (a)–(g).

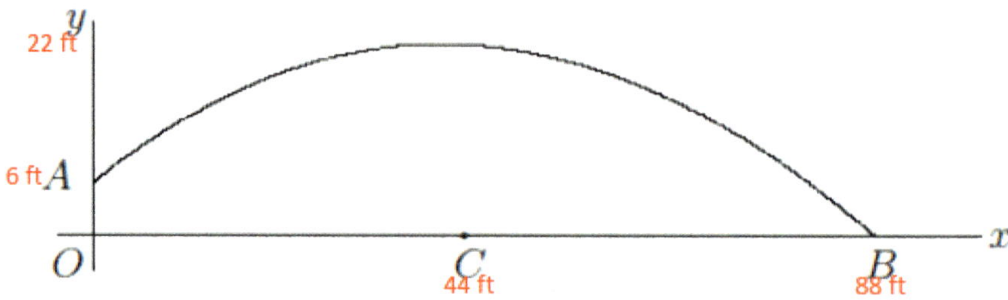

a. Suppose point A is approximately 6 feet above ground and that at time $t = 0$ the ball is at point A. Suppose the length of OB is approximately 88 feet. Include this information on the diagram.

   Information is noted on the diagram in red.

b. Suppose that after 1 second, the ball is at its highest point of 22 feet (above point C) and has traveled a horizontal distance of 44 feet. What are the approximate coordinates of the ball at the following values of t: $0.25, 0.5, 0.75, 1, 1.25, 1.5, 1.75$, and 2.

   *Most answers will vary because students are approximating the coordinates. The coordinates that must be correct because enough information was provided are denoted by a *.*

   At $t = 0.25$, the coordinates are approximately $(11, 10)$.

   At $t = 0.5$, the coordinates are approximately $(22, 18)$.

   At $t = 0.75$, the coordinates are approximately $(33, 20)$.

   **At $t = 1$, the coordinates are approximately $(44, 22)$.*

   At $t = 1.25$, the coordinates are approximately $(55, 19)$.

   At $t = 1.5$, the coordinates are approximately $(66, 14)$.

   At $t = 1.75$, the coordinates are approximately $(77, 8)$.

   **At $t = 2$, the coordinates are approximately $(88, 0)$.*

c. Use your answer from part (b) to write two predictions.

   Sample predictions:

   At a distance of 44 feet from where the ball was thrown, it is 22 feet in the air. At a distance of 66 feet from where the ball was thrown, it is 14 feet in the air.

d. What is happening to the ball when it has coordinates $(88, 0)$?

   At point $(88, 0)$, the ball has traveled for 2 seconds and has hit the ground at a distance of 88 feet from where the ball began.

   Lesson 1:   The Concept of a Function

This work is derived from Eureka Math ™ and licensed by Great Minds. ©2015 Great Minds. eureka-math.org
G8-M5-TE-B4-1.3.0-07.2015

e.   **Why do you think the ball is at point $(0, 6)$ when $t = 0$? In other words, why isn't the height of the ball 0?**

The ball is thrown from point A to point B. The fact that the ball is at a height of 6 feet means that the person throwing it must have released the ball from a height of 6 feet.

f.   **Does the graph allow us to make predictions about the height of the ball at all points?**

While we cannot predict exactly, the graph allows us to make approximate predictions of the height for any value of horizontal distance we choose.

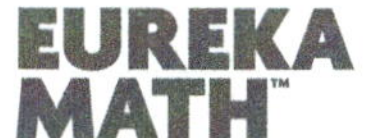

Lesson 1:   The Concept of a Function

This work is derived from Eureka Math ™ and licensed by Great Minds. ©2015 Great Minds. eureka-math.org
G8-M5-TE-B4-1.3.0-07.2015

Lesson 2: Formal Definition of a Function

Student Outcomes

- Students refine their understanding of the definition of a function.
- Students recognize that some, but not all, functions can be described by an equation between two variables.

Lesson Notes

A *function* is a correspondence between a set (whose elements are called *inputs*) and another set (whose elements are called *outputs*) such that each input corresponds to one and only one output. The correspondence is often given as a *rule* (e.g., the output is a number found by substituting an input number into the variable of a one-variable expression and evaluating). Students develop here their intuitive definition of a function as a rule that assigns to each element of one set of objects one, and only one, element from a second set of objects. Refinement of this definition, function notation, and detailed attention to the domain and range of functions are all left to the high school work of standards **F-IF.A.1** and **F-IF.B.5**.

We begin this lesson by looking at a troublesome set of data values.

Classwork

Opening (3 minutes)

- Shown below is the table from Example 2 of the last lesson and another table of values for the alleged motion of a second moving object. Make some comments about any troublesome features you observe in the second table of values. Does the first table of data have these troubles too?

Number of seconds (x)	Distance traveled in feet (y)
0.5	4
1	16
1.5	36
2	64
2.5	100
3	144
3.5	196
4	256

Number of seconds (x)	Distance traveled in feet (y)
0.5	4
1	4
1	36
2	64
2.5	80
3	99
3	196
4	256

Allow students to share their thoughts about the differences between the two tables. Then proceed with the discussion that follows.

    Lesson 2:   Formal Definition of a Function

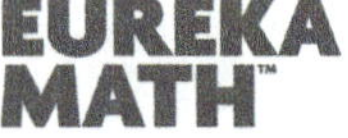

This work is derived from Eureka Math ™ and licensed by Great Minds. ©2015 Great Minds. eureka-math.org
G8-M5-TE-B4-1.3.0-07.2015

Discussion (8 minutes)

- Consider the object following the motion described on the left table. How far did it travel during the first second?

    □ *After 1 second, the object traveled 16 feet.*

- Consider the object following the motion described in the right table. How far did it travel during the first second?

    □ *It is unclear. After 1 second, the table indicates that the object traveled 4 feet and it also indicates that it traveled 36 feet.*

- Which of the two tables above allows us to make predictions with some accuracy? Explain.

    □ *The table on the left seems like it would be more accurate. The table on the right gives two completely different distances for the stone after 1 second. We cannot make an accurate prediction because after 1 second, the stone may either be 4 feet from where it started or 36 feet.*

- In the last lesson we defined a *function* to be a rule that assigns to each value of one quantity one, and only one, value of a second quantity. The right table does not follow this definition: it is assigning two different values, 4 feet and 36 feet, to the same time of 1 second. For the sake of meaningful discussion in a real-world situation this is problematic.

- Let's formalize this idea of assignment for the example of a falling stone from the last lesson. It seems more natural to use the symbol D, for distance, for the function that assigns to each time the distance the object has fallen by that time. So here D is a rule that assigns to each number t (with $0 \leq t \leq 4$) another number, the distance of the fall of the stone in t seconds. Here is the table from the last lesson.

Number of seconds (t)	Distance traveled in feet (D)
0.5	4
1	16
1.5	36
2	64
2.5	100
3	144
3.5	196
4	256

- We can interpret this table explicitly as a function rule:

D assigns the value 4 to the value 0.5.
D assigns the value 16 to the value 1.
D assigns the value 36 to the value 1.5.
D assigns the value 64 to the value 2.
D assigns the value 100 to the value 2.5.
D assigns the value 144 to the value 3.
D assigns the value 196 to the value 3.5.
D assigns the value 256 to the value 4.

Lesson 2:   Formal Definition of a Function

23

This work is derived from Eureka Math ™ and licensed by Great Minds. ©2015 Great Minds. eureka-math.org
G8-M5-TE-B4-1.3.0-07.2015

- If you like, you can think of this as an *input–output machine*. That is, we put in a number for the time (the input), and out comes another number (the output) that tells us the distance traveled in feet up to that time.

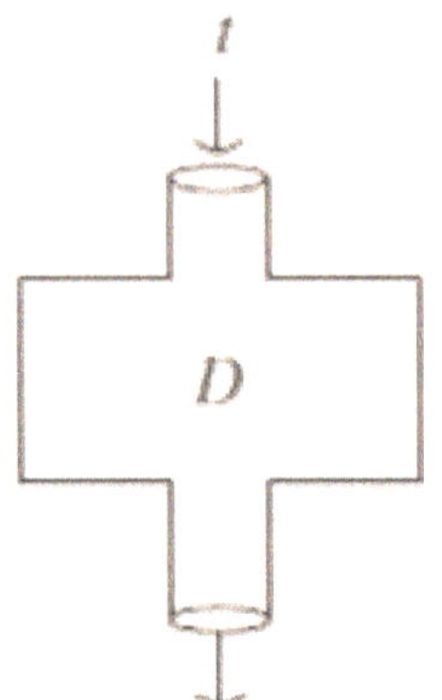

Distance traveled in t seconds

Scaffolding:

Highlighting the components of the words *input* and *output* and exploring how the words describe related concepts would be useful.

- With the example of the falling stone, what are we inputting?
    - *The input would be the time, in seconds, between 0 and 4 seconds.*
- What is the output?
    - *The output is the distance, in feet, the stone traveled up to that time.*
- If we input 3 into the machine, what is the output?
    - *The output is 144.*
- If we input 1.5 into the machine, what is the output?
    - *The output is 36.*
- Of course, with this particular machine, we are limited to inputs in the range of 0 to 4 because we are inputting times t during which the stone was falling.
- We are lucky with the function D: Sir Isaac Newton (1643–1727) studied the motion of objects falling under gravity and established a formula for their motion. It is given by $D = 16t^2$, that is the distance traveled over time interval t is $16t^2$. We can see that it fits our data values. Not all functions have equations describing them.

Time in seconds	1	2	3	4
Distance stone fell by that time in feet	16	64	144	256

- Functions can be represented in a variety of ways. At this point, we have seen the function that describes the distance traveled by the stone pictorially (from Lesson 1, Example 2), as a table of values, and as a rule described in words or as a mathematical equation.

Lesson 2:   Formal Definition of a Function

This work is derived from Eureka Math ™ and licensed by Great Minds. ©2015 Great Minds. eureka-math.org
G8-M5-TE-B4-1.3.0-07.2015

Exercise 1 (5 minutes)

Have students verify that $D = 16t^2$ does indeed match the data values of Example 1 by completing this next exercise. To expedite the verification, allow the use of calculators.

Exercises 1–5

1. Let D be the distance traveled in time t. Use the equation $D = 16t^2$ to calculate the distance the stone dropped for the given time t.

Time in seconds	0.5	1	1.5	2	2.5	3	3.5	4
Distance stone fell in feet by that time	4	16	36	64	100	144	196	256

  a. Are the distances you calculated equal to the table from Lesson 1?

    Yes

  b. Does the function $D = 16t^2$ accurately represent the distance the stone fell after a given time t? In other words, does the function described by this rule assign to t the correct distance? Explain.

    Yes, the function accurately represents the distance the stone fell after the given time interval. Each computation using the function resulted in the correct distance. Therefore, the function assigns to t the correct distance.

Discussion (10 minutes)

- Being able to write a formula for the function has superb implications—it is predictive. That is, we can predict what will happen each time a stone is released from a height of 256 feet. The equation describing the function makes it possible for us to know exactly how many feet the stone will fall for a time t as long as we select a t so that $0 \le t \le 4$.

- Not every function can be expressed as a formula, however. For example, consider the function H which assigns to each moment since you were born your height at that time. This is a function (Can you have two different heights at the same moment?), but it is very unlikely that there is a formula detailing your height over time.

- A function is a rule that assigns to each value of one quantity *exactly one* value of a second quantity. A *function* is a correspondence between a set of inputs and a set of outputs such that each input corresponds to one and only one output.

Note: Sometimes the phrase *exactly one* is used instead of *one and only one*. Both phrases mean the same thing; that is, an input with no corresponding output is unacceptable, and an input corresponding to several outputs is also unacceptable.

- Let's examine the definition of function more closely: For every input, there is *one and only one* output. Can you think of why the phrase *one and only one* (or *exactly one*) must be included in the definition?

   ▫ *We don't want an input-output machine that gives different output each time you put in the same input.*

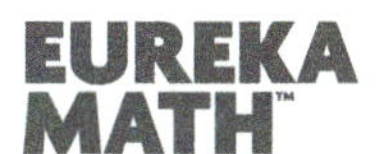

This work is derived from Eureka Math ™ and licensed by Great Minds. ©2015 Great Minds. eureka-math.org
G8-M5-TE-B4-1.3.0-07.2015

- Most of the time in Grade 8, the correspondence is given by a rule, which can also be considered a set of instructions used to determine the output for any given input. For example, a common rule is to substitute a number into the variable of a one-variable expression and evaluating. When a function is given by such a rule or formula, we often say that function is a rule that assigns to each input exactly one output.

- Is it clear that our function D, the rule that assigns to each time t satisfying $0 \leq t \leq 4$ the distance the object has fallen by that time, satisfies this condition of being a function?

Provide time for students to consider the phrase. Allow them to talk in pairs or small groups perhaps and then share their thoughts with the class. Use the question below, if necessary. Then resume the discussion.

- Using our stone-dropping example, if D assigns 64 to 2—that is, the function assigns 64 feet to the time 2 seconds—would it be possible for D to assign 65 to 2 as well? Explain.

  □ *It would not be possible for D to assign 64 and 65 to 2. The reason is that we are talking about a stone dropping. How could the stone drop 64 feet in 2 seconds **and** 65 feet in 2 seconds? The stone cannot be in two places at once.*

MP.6

- When given a formula for a function, we need to be careful of its context. For example, with our falling stone we have the formula $D = 16t^2$ describing the function. This formula holds for all values of time t with $0 \leq t \leq 4$. But it is also possible to put the value $t = -2$ into this formula and compute a supposed value of D:

$$D = 16(-2)^2$$
$$= 16(4)$$
$$= 64$$

Does this mean that for the two seconds before the stone was dropped it had fallen 64 feet? Of course not. We could also compute, for $t = 5$:

$$D = 16(5)^2$$
$$= 16(25)$$
$$= 400$$

- What is wrong with this statement?

  □ *It would mean that the stone dropped 400 feet in 5 seconds, but the stone was dropped from a height of 256 feet. It makes no sense.*

- To summarize, a function is a rule that assigns to each value of one quantity (an input) exactly one value to a second quantity (the matching output). Additionally, we should always consider the context when working with a function to make sure our answers makes sense: If a function is described by a formula, then we can only consider values to insert into that formula relevant to the context.

Lesson 2:   Formal Definition of a Function

This work is derived from Eureka Math ™ and licensed by Great Minds. ©2015 Great Minds. eureka-math.org
G8-M5-TE-B4-1.3.0-07.2015

Exercises 2–5 (10 minutes)

Students work independently to complete Exercises 2–5.

2. Can the table shown below represent values of a function? Explain.

Input (x)	1	3	5	5	9
Output (y)	7	16	19	20	28

No, the table cannot represent a function because the input of 5 has two different outputs. Functions assign only one output to each input.

3. Can the table shown below represent values of a function? Explain.

Input (x)	0.5	7	7	12	15
Output (y)	1	15	10	23	30

No, the table cannot represent a function because the input of 7 has two different outputs. Functions assign only one output to each input.

4. Can the table shown below represent values of a function? Explain.

Input (x)	10	20	50	75	90
Output (y)	32	32	156	240	288

Yes, the table can represent a function. Even though there are two outputs that are the same, each input has only one output.

5. It takes Josephine 34 minutes to complete her homework assignment of 10 problems. If we assume that she works at a constant rate, we can describe the situation using a function.

   a. Predict how many problems Josephine can complete in 25 minutes.

      Answers will vary.

   b. Write the two-variable linear equation that represents Josephine's constant rate of work.

      Let y be the number of problems she can complete in x minutes.

$$\frac{10}{34} = \frac{y}{x}$$

$$y = \frac{10}{34}x$$

$$y = \frac{5}{17}x$$

This work is derived from Eureka Math ™ and licensed by Great Minds. ©2015 Great Minds. eureka-math.org
G8-M5-TE-B4-1.3.0-07.2015

c. Use the equation you wrote in part (b) as the formula for the function to complete the table below. Round your answers to the hundredths place.

Time taken to complete problems (x)	5	10	15	20	25
Number of problems completed (y)	1.47	2.94	4.41	5.88	7.35

After 5 minutes, Josephine was able to complete 1.47 problems, which means that she was able to complete 1 problem, then get about halfway through the next problem.

d. Compare your prediction from part (a) to the number you found in the table above.

Answers will vary.

e. Use the formula from part (b) to compute the number of problems completed when $x = -7$. Does your answer make sense? Explain.

$$y = \frac{5}{17}(-7)$$
$$= -2.06$$

No, the answer does not make sense in terms of the situation. The answer means that Josephine can complete -2.06 problems in -7 minutes. This obviously does not make sense.

f. For this problem, we assumed that Josephine worked at a constant rate. Do you think that is a reasonable assumption for this situation? Explain.

It does not seem reasonable to assume constant rate for this situation. Just because Josephine was able to complete 10 problems in 34 minutes does not necessarily mean she spent the exact same amount of time on each problem. For example, it may have taken her 20 minutes to do 1 problem and then 14 minutes total to finish the remaining 9 problems.

Closing (4 minutes)

Summarize, or ask students to summarize, the main points from the lesson:

- We know that a function is a rule that assigns to each value of one quantity (an input) exactly one value of a second quantity (its matching output).
- Not every function can be described by a mathematical formula.
- If we can describe a function by a mathematical formula, we must still be careful of context. For example, asking for the distance a stone drops in -2 seconds is meaningless.

Lesson 2:   Formal Definition of a Function

EUREKA MATH

This work is derived from Eureka Math ™ and licensed by Great Minds. ©2015 Great Minds. eureka-math.org
G8-M5-TE-B4-1.3.0-07.2015

Lesson Summary

A *function* is a correspondence between a set (whose elements are called *inputs*) and another set (whose elements are called *outputs*) such that each input corresponds to one and only one output.

Sometimes the phrase *exactly one output* is used instead of *one and only one output* in the definition of function (they mean the same thing). Either way, it is this fact, that there is one and only one output for each input, which makes functions predictive when modeling real life situations.

Furthermore, the correspondence in a function is often given by a *rule* (or *formula*). For example, the output is equal to the number found by substituting an input number into the variable of a one-variable expression and evaluating.

Functions are sometimes described as an *input–output machine*. For example, given a function D, the input is time t, and the output is the distance traveled in t seconds.

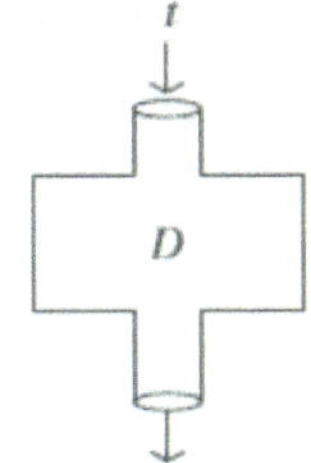

Distance traveled in t seconds

Exit Ticket (5 minutes)

This work is derived from Eureka Math ™ and licensed by Great Minds. ©2015 Great Minds. eureka-math.org
G8-M5-TE-B4-1.3.0-07.2015

Name ____   Date ___________________

Lesson 2: Formal Definition of a Function

Exit Ticket

1. Can the table shown below represent values of a function? Explain.

Input (x)	10	20	30	40	50
Output (y)	32	64	96	64	32

2. Kelly can tune 4 cars in 3 hours. If we assume he works at a constant rate, we can describe the situation using a function.

   a. Write the function that represents Kelly's constant rate of work.

   b. Use the function you wrote in part (a) as the formula for the function to complete the table below. Round your answers to the hundredths place.

Time spent tuning cars (x)	2	3	4	6	7
Number of cars tuned up (y)					

This work is derived from Eureka Math ™ and licensed by Great Minds. ©2015 Great Minds. eureka-math.org
G8-M5-TE-B4-1.3.0-07.2015

EUREKA MATH

c.   Kelly works 8 hours per day. According to this work, how many cars will he finish tuning at the end of a shift?

d.   For this problem, we assumed that Kelly worked at a constant rate. Do you think that is a reasonable assumption for this situation? Explain.

This work is derived from Eureka Math ™ and licensed by Great Minds. ©2015 Great Minds. eureka-math.org
G8-M5-TE-B4-1.3.0-07.2015

Exit Ticket Sample Solutions

1. Can the table shown below represent values of a function? Explain.

Input (x)	10	20	30	40	50
Output (y)	32	64	96	64	32

Yes, the table can represent a function. Each input has exactly one output.

2. Kelly can tune 4 cars in 3 hours. If we assume he works at a constant rate, we can describe the situation using a function.

   a. Write the function that represents Kelly's constant rate of work.

   Let y represent the number of cars Kelly can tune up in x hours; then

   $$\frac{y}{x} = \frac{4}{3}$$
   $$y = \frac{4}{3}x$$

   b. Use the function you wrote in part (a) as the formula for the function to complete the table below. Round your answers to the hundredths place.

Time spent tuning cars (x)	2	3	4	6	7
Number of cars tuned up (y)	2.67	4	5.33	8	9.33

   c. Kelly works 8 hours per day. According to this work, how many cars will he finish tuning at the end of a shift?

   Using the function, Kelly will tune up 10.67 cars at the end of his shift. That means he will finish tuning up 10 cars and begin tuning up the 11th car.

   d. For this problem, we assumed that Kelly worked at a constant rate. Do you think that is a reasonable assumption for this situation? Explain.

   No, it does not seem reasonable to assume a constant rate for this situation. Just because Kelly tuned up 4 cars in 3 hours does not mean he spent the exact same amount of time on each car. One car could have taken 1 hour, while the other three could have taken 2 hours total.

Lesson 2:   Formal Definition of a Function

EUREKA MATH™

This work is derived from Eureka Math ™ and licensed by Great Minds. ©2015 Great Minds. eureka-math.org
G8-M5-TE-B4-1.3.0-07.2015

Problem Set Sample Solutions

1. The table below represents the number of minutes Francisco spends at the gym each day for a week. Does the data shown below represent values of a function? Explain.

Day (x)	1	2	3	4	5	6	7
Time in minutes (y)	35	45	30	45	35	0	0

Yes, the table can represent a function because each input has a unique output. For example, on day 1, Francisco was at the gym for 35 minutes.

2. Can the table shown below represent values of a function? Explain.

Input (x)	9	8	7	8	9
Output (y)	11	15	19	24	28

No, the table cannot represent a function because the input of 9 has two different outputs, and so does the input of 8. Functions assign only one output to each input.

3. Olivia examined the table of values shown below and stated that a possible rule to describe this function could be $y = -2x + 9$. Is she correct? Explain.

Input (x)	-4	0	4	8	12	16	20	24
Output (y)	17	9	1	-7	-15	-23	-31	-39

Yes, Olivia is correct. When the rule is used with each input, the value of the output is exactly what is shown in the table. Therefore, the rule for this function could well be $y = -2x + 9$.

4. Peter said that the set of data in part (a) describes a function, but the set of data in part (b) does not. Do you agree? Explain why or why not.

a.

Input (x)	1	2	3	4	5	6	7	8
Output (y)	8	10	32	6	10	27	156	4

b.

Input (x)	-6	-15	-9	-3	-2	-3	8	9
Output (y)	0	-6	8	14	1	2	11	41

Peter is correct. The table in part (a) fits the definition of a function. That is, there is exactly one output for each input. The table in part (b) cannot be a function. The input -3 has two outputs, 14 and 2. This contradicts the definition of a function; therefore, it is not a function.

This work is derived from Eureka Math ™ and licensed by Great Minds. ©2015 Great Minds. eureka-math.org
G8-M5-TE-B4-1.3.0-07.2015

5. A function can be described by the rule $y = x^2 + 4$. Determine the corresponding output for each given input.

Input (x)	-3	-2	-1	0	1	2	3	4
Output (y)	13	8	5	4	5	8	13	20

6. Examine the data in the table below. The inputs and outputs represent a situation where constant rate can be assumed. Determine the rule that describes the function.

Input (x)	-1	0	1	2	3	4	5	6
Output (y)	3	8	13	18	23	28	33	38

The rule that describes this function is $y = 5x + 8$.

7. Examine the data in the table below. The inputs represent the number of bags of candy purchased, and the outputs represent the cost. Determine the cost of one bag of candy, assuming the price per bag is the same no matter how much candy is purchased. Then, complete the table.

Bags of Candy (x)	1	2	3	4	5	6	7	8
Cost in Dollars (y)	1.25	2.50	3.75	5.00	6.25	7.50	8.75	10.00

a. Write the rule that describes the function.

$y = 1.25x$

b. Can you determine the value of the output for an input of $x = -4$? If so, what is it?

When $x = -4$, the output is -5.

c. Does an input of -4 make sense in this situation? Explain.

No, an input of -4 does not make sense for the situation. It would mean -4 bags of candy. You cannot purchase -4 bags of candy.

8. Each and every day a local grocery store sells 2 pounds of bananas for $\$1.00$. Can the cost of 2 pounds of bananas be represented as a function of the day of the week? Explain.

Yes, this situation can be represented by a function. Assign to each day of the week the value $\$1.00$.

EUREKA MATH™

This work is derived from Eureka Math ™ and licensed by Great Minds. ©2015 Great Minds. eureka-math.org
G8-M5-TE-B4-1.3.0-07.2015

9. Write a brief explanation to a classmate who was absent today about why the table in part (a) is a function and the table in part (b) is not.

a.

Input (x)	-1	-2	-3	-4	4	3	2	1
Output (y)	81	100	320	400	400	320	100	81

b.

Input (x)	1	6	-9	-2	1	-10	8	14
Output (y)	2	6	-47	-8	19	-2	15	31

The table in part (a) is a function because each input has exactly one output. This is different from the information in the table in part (b). Notice that the input of 1 has been assigned two different values. The input of 1 is assigned 2 and 19. Because the input of 1 has more than one output, this table cannot represent a function.

This work is derived from Eureka Math ™ and licensed by Great Minds. ©2015 Great Minds. eureka-math.org
G8-M5-TE-B4-1.3.0-07.2015

Lesson 3: Linear Functions and Proportionality

Student Outcomes

- Students realize that linear equations of the form $y = mx + b$ can be seen as rules defining functions (appropriately called *linear functions*).
- Students explore examples of linear functions.

Classwork

Example 1 (7 minutes)

Example 1

In the last lesson, we looked at several tables of values showing the inputs and outputs of functions. For instance, one table showed the costs of purchasing different numbers of bags of candy:

Bags of candy (x)	1	2	3	4	5	6	7	8
Cost in Dollars (y)	1.25	2.50	3.75	5.00	6.25	7.50	8.75	10.00

- What do you think a *linear* function is?
  - *A linear function is likely a function with a rule described by a linear equation. Specifically, the rate of change in the situation being described is constant, and the graph of the equation is a line.*

- Do you think this is a *linear* function? Justify your answer.
  - *Yes, this is a linear function because there is a proportional relationship:*

    $$\frac{10.00}{8} = 1.25; \ \$1.25 \text{ per each bag of candy}$$

    $$\frac{5.00}{4} = 1.25; \ \$1.25 \text{ per each bag of candy}$$

    $$\frac{2.50}{2} = 1.25; \ \$1.25 \text{ per each bag of candy}$$

> **Scaffolding:**
>
> In addition to explanations about functions, it may be useful for students to have a series of structured experiences with real-world objects and data to reinforce their understanding of a function. An example is experimenting with different numbers of batches of a given recipe; students can observe the effect of the number of batches on quantities of various ingredients.

Lesson 3:   Linear Functions and Proportionality

EUREKA MATH™

This work is derived from Eureka Math ™ and licensed by Great Minds. ©2015 Great Minds. eureka-math.org
G8-M5-TE-B4-1.3.0-07.2015

□   *The total cost is increasing at a rate of $1.25 with each bag of candy. Further justification comes from the graph of the data shown below.*

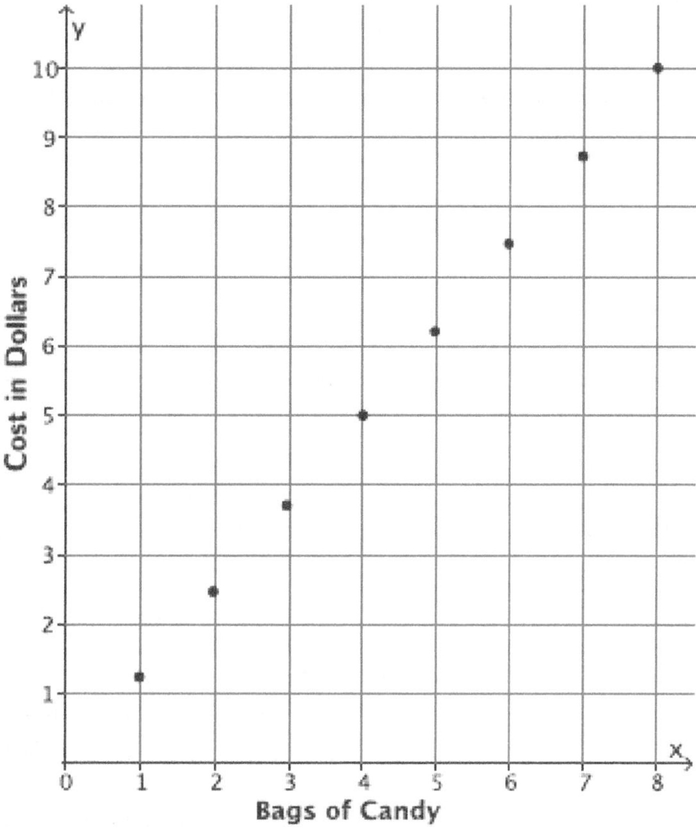

- A *linear function* is a function with a rule that can be described by a linear equation. That is, if we use x to denote an input of the function and y its matching output, then the function is linear if the rule for the function can be described by the equation $y = mx + b$ for some numbers m and b.

- What rule or equation describes our cost function for bags of candy?

  □   *The rule that represents the function is then $y = 1.25x$.*

- Notice that the constant m is 1.25, which is the cost of one bag of candy, and the constant b is 0. Also notice that the constant m was found by calculating the unit rate for a bag of candy.

- No matter the value of x chosen, as long as x is a nonnegative integer, the rule $y = 1.25x$ gives the cost of purchasing that many bags of candy. *The total cost of candy is a function of the number of bags purchased.*

- Why must we set x as a nonnegative integer for this function?

  □   *Since x represents the number of bags of candy, it does not make sense that there would be a negative number of bags. It is also unlikely that we might be allowed to buy fractional bags of candy, and so we require x to be a whole number.*

- Would you say that the table represents all possible inputs and outputs? Explain.

  □   *No, it does not represent all possible inputs and outputs. Someone can purchase more than 8 bags of candy, and inputs greater than 8 are not represented by this table (unless the store has a limit on the number of bags one may purchase, perhaps).*

Lesson 3:   Linear Functions and Proportionality

This work is derived from Eureka Math ™ and licensed by Great Minds. ©2015 Great Minds. eureka-math.org
G8-M5-TE-B4-1.3.0-07.2015

- As a matter of precision we say that "this function has the above table of values" instead of "the table above represents a function" because not all values of the function might be represented by the table. Any rule that describes a function usually applies to all of the possible values of a function. For example, in our candy example, we can determine the cost for 9 bags of candy using our equation for the function even though no column is shown in the table for when x is 9. However, if for some reason our interest was in a function with only the input values $1, 2, 3, 4, 5, 6, 7,$ and 8, then our table gives complete information about the function and so fully represents the function.

Example 2 (4 minutes)

> **Example 2**
>
> Walter walks at a constant speed of 8 miles every 2 hours. Describe a linear function for the number of miles he walks in x hours. What is a reasonable range of x-values for this function?

- Consider the following rate problem: Walter walks at a constant speed of 8 miles every 2 hours. Describe a linear function for the number of miles he walks in x hours. What is a reasonable range of x-values for this function?

  - *Walter's average speed of walking 8 miles is* $\dfrac{8}{2} = 4$, *or 4 miles per hour.*

- We have $y = 4x$, where y is the distance walked in x hours. It seems reasonable to say that x is any real number between 0 and 20, perhaps? (Might there be a cap on the number of hours he walks? Perhaps we are counting up the number of miles he walks over a lifetime?)

  > *Scaffolding:*
  >
  > As the language becomes more abstract, it can be useful to use visuals and even pantomime situations related to speed, rate, etc.

- In the last example, the total cost of candy was a function of the number of bags purchased. Describe the function in this walking example.

  - *The distance that Walter travels is a function of the number of hours he spends walking.*

- What limitations did we put on x?

  > *We must insist that* $x \geq 0$. *Since x represents the time Walter walks, then it makes sense that he would walk for a positive amount of time or no time at all.*

- Since x is positive, then we know that the distance y will also be positive.

Example 3 (4 minutes)

> **Example 3**
>
> Veronica runs at a constant speed. The distance she runs is a function of the time she spends running. The function has the table of values shown below.
>
> | Time in minutes (x) | 8 | 16 | 24 | 32 |
> |---|---|---|---|---|
> | Distance run in miles (y) | 1 | 2 | 3 | 4 |

Lesson 3:   Linear Functions and Proportionality

EUREKA MATH

This work is derived from Eureka Math ™ and licensed by Great Minds. ©2015 Great Minds. eureka-math.org
G8-M5-TE-B4-1.3.0-07.2015

- Since Veronica runs at a constant speed, we know that her average speed over any time interval will be the same. Therefore, Veronica's distance function is a linear function. Write the equation that describes her distance function.

    □ *The function that represents Veronica's distance is described by the equation* $y = \frac{1}{8}x$*, where* y *is the distance in miles Veronica runs in* x *minutes and* $x, y \geq 0$.

- Describe the function in terms of distance and time.

    □ *The distance that Veronica runs is a function of the number of minutes she spends running.*

Example 4 (5 minutes)

Example 4

Water flows from a faucet into a bathtub at the constant rate of 7 gallons of water pouring out every 2 minutes. The bathtub is initially empty, and its plug is in. Determine the rule that describes the volume of water in the tub as a function of time. If the tub can hold 50 gallons of water, how long will it take to fill the tub?

The rate of water flow is $\frac{7}{2}$*, or* 3.5 *gallons per minute. Then the rule that describes the volume of water in the tub as a function of time is* $y = 3.5x$*, where* y *is the volume of water, and* x *is the number of minutes the faucet has been on.*

To find the time when $y = 50$*, we need to look at the equation* $50 = 3.5x$*. This gives* $x = \frac{50}{3.5} = 14.2857\ldots \approx 14$*. It will take about* 14 *minutes to fill the tub.*

- Assume that the faucet is filling a bathtub that can hold 50 gallons of water. How long will it take the faucet to fill the tub?

    □ *Since we want the total volume to be* 50 *gallons, then*

$$50 = 3.5x$$

$$\frac{50}{3.5} = x$$

$$14.2857\ldots = x$$

$$14 \approx x$$

    It will take about 14 *minutes to fill a tub that has a volume of* 50 *gallons.*

Now assume that you are filling the same 50-gallon bathtub with water flowing in at the constant rate of 3.5 gallons per minute, but there were initially 8 gallons of water in the tub. Will it still take about 14 minutes to fill the tub?

No. It will take less time because there is already some water in the tub.

- What now is the appropriate equation describing the volume of water in the tub as a function of time?

    □ *If* y *is the volume of water that flows from the faucet, and* x *is the number of minutes the faucet has been on, then* $y = 3.5x + 8$.

Lesson 3:   Linear Functions and Proportionality

39

This work is derived from Eureka Math ™ and licensed by Great Minds. ©2015 Great Minds. eureka-math.org
G8-M5-TE-B4-1.3.0-07.2015

- How long will it take to fill the tub according to this equation?
  - *Since we still want the total volume of the tub to be* 50 *gallons, then:*

$$50 = 3.5x + 8$$
$$42 = 3.5x$$
$$12 = x$$

  It will take 12 *minutes for the faucet to fill a* 50-*gallon tub when* 8 *gallons are already in it.*

  (Be aware that some students may observe that we can use the previous function rule $y = 3.5x$ to answer this question by noting that we need to add only 42 more gallons to the tub. This will lead directly to the equation $42 = 3.5x$.)

- Generate a table of values for this function:

Time in minutes (x)	0	3	6	9	12
Total volume in tub in gallons (y)	8	18.5	29	39.5	50

Example 5 (7 minutes)

Example 5

Water flows from a faucet at a constant rate. Assume that 6 gallons of water are already in a tub by the time we notice the faucet is on. This information is recorded in the first column of the table below. The other columns show how many gallons of water are in the tub at different numbers of minutes since we noticed the running faucet.

Time in minutes (x)	0	3	5	9
Total volume in tub in gallons (y)	6	9.6	12	16.8

- After 3 minutes pass, there are 9.6 gallons in the tub. How much water flowed from the faucet in those 3 minutes? Explain.
  - *Since there were already* 6 *gallons in the tub, after* 3 *minutes an additional* 3.6 *gallons filled the tub.*
- Use this information to determine the rate of water flow.
  - *In* 3 *minutes,* 3.6 *gallons were added to the tub, then* $\dfrac{3.6}{3} = 1.2$, *and the faucet fills the tub at a rate of* 1.2 *gallons per minute.*
- Verify that the rate of water flow is correct using the other values in the table.
  - *Sample student work:*

    $5(1.2) = 6$, *and since* 6 *gallons were already in the tub, the total volume in the tub is* 12 *gallons.*

    $9(1.2) = 10.8$, *and since* 6 *gallons were already in the tub, the total volume in the tub is* 16.8 *gallons.*
- Write an equation that describes the volume of water in the tub as a function of time.
  - *The volume function that represents the rate of water flow from the faucet is* $y = 1.2x + 6$, *where* y *is the volume of water in the tub, and* x *is the number of minutes that have passed since we first noticed the faucet being on.*

Lesson 3:   Linear Functions and Proportionality

This work is derived from Eureka Math ™ and licensed by Great Minds. ©2015 Great Minds. eureka-math.org
G8-M5-TE-B4-1.3.0-07.2015

▪ For how many minutes was the faucet on before we noticed it? Explain.

> *Since* 6 *gallons were in the tub by the time we noticed the faucet was on, we need to determine how many minutes it takes for* 6 *gallons to flow into the tub. The columns for* $x = 0$ *and* $x = 5$ *in the table show that six gallons of water pour in the tub over a five-minute period. The faucet was on for* 5 *minutes before we noticed it.*

Exercises 1–3 (10 minutes)

Students complete Exercises 1–3 independently or in pairs.

Exercises 1–3

1. Hana claims she mows lawns at a constant rate. The table below shows the area of lawn she can mow over different time periods.

Number of minutes (x)	5	20	30	50
Area mowed in square feet (y)	36	144	216	360

a. Is the data presented consistent with the claim that the area mowed is a linear function of time?

Sample responses:

Linear functions have a constant rate of change. When we compare the rates at each interval of time, they will be equal to the same constant.

When the data is graphed on the coordinate plane, it appears to make a line.

b. Describe in words the function in terms of area mowed and time.

The total area mowed is a function of the number of minutes spent mowing.

c. At what rate does Hana mow lawns over a 5-minute period?

$$\frac{36}{5} = 7.2$$

The rate is 7.2 *square feet per minute.*

d. At what rate does Hana mow lawns over a 20-minute period?

$$\frac{144}{20} = 7.2$$

The rate is 7.2 *square feet per minute.*

e. At what rate does Hana mow lawns over a 30-minute period?

$$\frac{216}{30} = 7.2$$

The rate is 7.2 *square feet per minute.*

This work is derived from Eureka Math ™ and licensed by Great Minds. ©2015 Great Minds. eureka-math.org
G8-M5-TE-B4-1.3.0-07.2015

f. At what rate does Hana mow lawns over a 50-minute period?

$$\frac{360}{50} = 7.2$$

The rate is 7.2 square feet per minute.

g. Write the equation that describes the area mowed, y, in square feet, as a linear function of time, x, in minutes.

$$y = 7.2x$$

h. Describe any limitations on the possible values of x and y.

Both x and y must be positive numbers. The symbol x represents time spent mowing, which means it should be positive. Similarly, y represents the area mowed, which should also be positive.

i. What number does the function assign to $x = 24$? That is, what area of lawn can be mowed in 24 minutes?

$$y = 7.2(24)$$
$$y = 172.8$$

In 24 minutes, an area of 172.8 square feet can be mowed.

j. According to this work, how many minutes would it take to mow an area of 400 square feet?

$$400 = 7.2x$$
$$\frac{400}{7.2} = x$$
$$55.555\ldots = x$$
$$56 \approx x$$

It would take about 56 minutes to mow an area of 400 square feet.

2. A linear function has the table of values below. The information in the table shows the total volume of water, in gallons, that flows from a hose as a function of time, the number of minutes the hose has been running.

Time in minutes (x)	10	25	50	70
Total volume of water in gallons (y)	44	110	220	308

a. Describe the function in terms of volume and time.

The total volume of water that flows from a hose is a function of the number of minutes the hose is left on.

b. Write the rule for the volume of water in gallons, y, as a linear function of time, x, given in minutes.

$$y = \frac{44}{10}x$$
$$y = 4.4x$$

Lesson 3:   Linear Functions and Proportionality

This work is derived from Eureka Math ™ and licensed by Great Minds. ©2015 Great Minds. eureka-math.org
G8-M5-TE-B4-1.3.0-07.2015

EUREKA MATH™

c. What number does the function assign to 250? That is, how many gallons of water flow from the hose during a period of 250 minutes?

$$y = 4.4(250)$$
$$y = 1\,100$$

In 250 minutes, 1,100 gallons of water flow from the hose.

d. The average swimming pool holds about 17,300 gallons of water. Suppose such a pool has already been filled one quarter of its volume. Write an equation that describes the volume of water in the pool if, at time 0 minutes, we use the hose described above to start filling the pool.

$$\frac{1}{4}(17\,300) = 4\,325$$
$$y = 4.4x + 4\,325$$

e. Approximately how many hours will it take to finish filling the pool?

$$17\,300 = 4.4x + 4\,325$$
$$12\,975 = 4.4x$$
$$\frac{12\,975}{4.4} = x$$
$$2\,948.8636\ldots = x$$
$$2\,949 \approx x$$

$$\frac{2\,949}{60} = 49.15$$

It will take about 49 hours to fill the pool with the hose.

3. Recall that a linear function can be described by a rule in the form of $y = mx + b$, where m and b are constants. A particular linear function has the table of values below.

Input (x)	0	4	10	11	15	20	23
Output (y)	4	24	54	59	79	104	119

a. What is the equation that describes the function?

$$y = 5x + 4$$

b. Complete the table using the rule.

This work is derived from Eureka Math ™ and licensed by Great Minds. ©2015 Great Minds. eureka-math.org
G8-M5-TE-B4-1.3.0-07.2015

Closing (4 minutes)

Summarize, or ask students to summarize, the main points from the lesson:

- We know that a linear function is a function whose rule can be described by a linear equation $y = mx + b$ with m and b as constants.
- We know that problems involving constant rates and proportional relationships can be described by linear functions.

> **Lesson Summary**
>
> A linear equation $y = mx + b$ describes a rule for a function. We call any function defined by a linear equation a linear function.
>
> Problems involving a constant rate of change or a proportional relationship can be described by linear functions.

Exit Ticket (4 minutes)

This work is derived from Eureka Math ™ and licensed by Great Minds. ©2015 Great Minds. eureka-math.org
G8-M5-TE-B4-1.3.0-07.2015

EUREKA MATH

Name ____   Date _________________

Lesson 3: Linear Functions and Proportionality

The information in the table shows the number of pages a student can read in a certain book as a function of time in minutes spent reading. Assume a constant rate of reading.

Time in minutes (x)	2	6	11	20
Total number of pages read in a certain book (y)	7	21	38.5	70

a.   Write the equation that describes the total number of pages read, y, as a linear function of the number of minutes, x, spent reading.

b.   How many pages can be read in 45 minutes?

c.   A certain book has 396 pages. The student has already read $\frac{3}{8}$ of the pages and now picks up the book again at time $x = 0$ minutes. Write the equation that describes the total number of pages of the book read as a function of the number of minutes of further reading.

d.   Approximately how much time, in minutes, will it take to finish reading the book?

Exit Ticket Sample Solutions

The information in the table shows the number of pages a student can read in a certain book as a function of time in minutes spent reading. Assume a constant rate of reading.

Time in minutes (x)	2	6	11	20
Total number of pages read in a certain book (y)	7	21	38.5	70

a. Write the equation that describes the total number of pages read, y, as a linear function of the number of minutes, x, spent reading.

$$y = \frac{7}{2}x$$
$$y = 3.5x$$

b. How many pages can be read in 45 minutes?

$$y = 3.5(45)$$
$$y = 157.5$$

In 45 minutes, the student can read 157.5 pages.

c. A certain book has 396 pages. The student has already read $\frac{3}{8}$ of the pages and now picks up the book again at time $x = 0$ minutes. Write the equation that describes the total number of pages of the book read as a function of the number of minutes of further reading.

$$\frac{3}{8}(396) = 148.5$$
$$y = 3.5x + 148.5$$

d. Approximately how much time, in minutes, will it take to finish reading the book?

$$396 = 3.5x + 148.5$$
$$247.5 = 3.5x$$
$$\frac{247.5}{3.5} = x$$
$$70.71428571\ldots = x$$
$$71 \approx x$$

It will take about 71 minutes to finish reading the book.

Lesson 3:   Linear Functions and Proportionality

EUREKA MATH

This work is derived from Eureka Math ™ and licensed by Great Minds. ©2015 Great Minds. eureka-math.org
G8-M5-TE-B4-1.3.0-07.2015

Problem Set Sample Solutions

1. A food bank distributes cans of vegetables every Saturday. The following table shows the total number of cans they have distributed since the beginning of the year. Assume that this total is a linear function of the number of weeks that have passed.

Number of weeks (x)	1	12	20	45
Number of cans of vegetables distributed (y)	180	2,160	3,600	8,100

a. Describe the function being considered in words.

The total number of cans handed out is a function of the number of weeks that pass.

b. Write the linear equation that describes the total number of cans handed out, y, in terms of the number of weeks, x, that have passed.

$$y = \frac{180}{1}x$$
$$y = 180x$$

c. Assume that the food bank wants to distribute $20{,}000$ cans of vegetables. How long will it take them to meet that goal?

$$20\,000 = 180x$$
$$\frac{20\,000}{180} = x$$
$$111.1111\ldots = x$$
$$111 \approx x$$

It will take about 111 weeks to distribute $20{,}000$ cans of vegetables, or about 2 years.

d. The manager had forgotten to record that they had distributed $35{,}000$ cans on January 1. Write an adjusted linear equation to reflect this forgotten information.

$$y = 180x + 35\,000$$

e. Using your function in part (d), determine how long in years it will take the food bank to hand out $80{,}000$ cans of vegetables.

$$80\,000 = 180x + 35\,000$$
$$45\,000 = 180x$$
$$\frac{45\,000}{180} = x$$
$$250 = x$$

To determine the number of years:

$$\frac{250}{52} = 4.8076\ldots \approx 4.8$$

It will take about 4.8 years to distribute $80{,}000$ cans of vegetables.

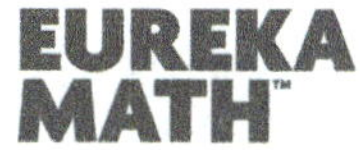

This work is derived from Eureka Math ™ and licensed by Great Minds. ©2015 Great Minds. eureka-math.org
G8-M5-TE-B4-1.3.0-07.2015

2. A linear function has the table of values below. It gives the number of miles a plane travels over a given number of hours while flying at a constant speed.

Number of hours traveled (x)	2.5	4	4.2
Distance in miles (y)	1,062.5	1,700	1,785

a. Describe in words the function given in this problem.

The total distance traveled is a function of the number of hours spent flying.

b. Write the equation that gives the distance traveled, y, in miles, as a linear function of the number of hours, x, spent flying.

$$y = \frac{1\,062.5}{2.5}x$$
$$y = 425x$$

c. Assume that the airplane is making a trip from New York to Los Angeles, which is a journey of approximately $2,475$ miles. How long will it take the airplane to get to Los Angeles?

$$2\,475 = 425x$$
$$\frac{2\,475}{425} = x$$
$$5.82352\ldots = x$$
$$5.8 \approx x$$

It will take about 5.8 hours for the airplane to fly $2,475$ miles.

d. If the airplane flies for 8 hours, how many miles will it cover?

$$y = 425(8)$$
$$y = 3\,400$$

The airplane would travel $3,400$ miles in 8 hours.

3. A linear function has the table of values below. It gives the number of miles a car travels over a given number of hours.

Number of hours traveled (x)	3.5	3.75	4	4.25
Distance in miles (y)	203	217.5	232	246.5

a. Describe in words the function given.

The total distance traveled is a function of the number of hours spent traveling.

b. Write the equation that gives the distance traveled, in miles, as a linear function of the number of hours spent driving.

$$y = \frac{203}{3.5}x$$
$$y = 58x$$

Lesson 3:   Linear Functions and Proportionality

EUREKA MATH™

This work is derived from Eureka Math ™ and licensed by Great Minds. ©2015 Great Minds. eureka-math.org
G8-M5-TE-B4-1.3.0-07.2015

c. Assume that the person driving the car is going on a road trip to reach a location 500 miles from her starting point. How long will it take the person to get to the destination?

$$500 = 58x$$

$$\frac{500}{58} = x$$

$$8.6206\ldots = x$$

$$8.6 \approx x$$

It will take about 8.6 hours to travel 500 miles.

4. A particular linear function has the table of values below.

Input (x)	2	3	8	11	15	20	23
Output (y)	7	10	25	34	46	61	70

a. What is the equation that describes the function?

$$y = 3x + 1$$

b. Complete the table using the rule.

5. A particular linear function has the table of values below.

Input (x)	0	5	8	13	15	18	21
Output (y)	6	11	14	19	21	24	27

a. What is the rule that describes the function?

$$y = x + 6$$

b. Complete the table using the rule.

This work is derived from Eureka Math ™ and licensed by Great Minds. ©2015 Great Minds. eureka-math.org
G8-M5-TE-B4-1.3.0-07.2015

Lesson 4: More Examples of Functions

Student Outcomes

- Students classify functions as either discrete or not discrete.

Classwork

Discussion (5 minutes)

- Consider two functions we discussed in previous lessons: the function that assigns to each count of candy bags the cost of purchasing that many bags (Table A) and the function that assigns to each time value between 0 and 4 seconds the distance a falling stone has dropped up to that time (Table B).

Table A:

Bags of candy (x)	1	2	3	4	5	6	7	8
Cost in Dollars (y)	1.25	2.50	3.75	5.00	6.25	7.50	8.75	10.00

Table B:

Number of seconds (x)	0.5	1	1.5	2	2.5	3	3.5	4
Distance traveled in feet (y)	4	16	36	64	100	144	196	256

- How do the two functions differ in the types of inputs they each admit?

Solicit answers from students, and then continue with the discussion below.

- We described the first function, Table A, with the rule $y = 1.25x$ for $x \geq 0$.
- Why did we restrict x to numbers equal to or greater than 0?
  - *We restricted x to numbers equal to or greater than 0 because you cannot purchase a negative number of bags of candy.*
- If we further assume that only a whole number of bags can be sold (that one cannot purchase only a portion of a bag's contents), then we need to be more precise about our restrictions on permissible inputs for the function. Specifically, we must say that x is a nonnegative integer: 0, 1, 2, 3, etc.
- We described the second function, Table B, with the rule $y = 16x^2$. Does this function require the same restrictions on its inputs as the previous problem? Explain.
  - *We should state that x must be a positive number because x represents the amount of time traveled. But we do not need to say that x must be an integer: intervals of time need not be in whole seconds as fractional counts of seconds are meaningful.*

Lesson 4:   More Examples of Functions

This work is derived from Eureka Math ™ and licensed by Great Minds. ©2015 Great Minds. eureka-math.org
G8-M5-TE-B4-1.3.0-07.2015

- The word *discrete* in English means individually separate or distinct. If a function admits only individually separate input values (like whole number counts of candy bags, for example), then we say we have *a discrete function*. If a function admits, over a range of values, <u>any</u> input value within that range (all fractional values too, for example), then we have a function that is not discrete. Functions that describe motion, for example, are typically not discrete.

Example 1 (6 minutes)

Practice the definitions with this example.

Example 1

Classify each of the functions described below as either discrete or not discrete.

a)   The function that assigns to each whole number the cost of buying that many cans of beans in a particular grocery store.

b)   The function that assigns to each time of day one Wednesday the temperature of Sammy's fever at that time.

c)   The function that assigns to each real number its first digit.

d)   The function that assigns to each day in the year 2015 my height at noon that day.

e)   The function that assigns to each moment in the year 2015 my height at that moment.

f)   The function that assigns to each color the first letter of the name of that color.

g)   The function that assigns the number 23 to each and every real number between 20 and 30.6.

h)   The function that assigns the word YES to every yes/no question.

i)   The function that assigns to each height directly above the North Pole the temperature of the air at that height right at this very moment.

a) Discrete   b) Not discrete   c) Not discrete   d) Discrete   e) Not discrete   f) Discrete   g) Not discrete

h) Discrete   i) Not discrete

Example (2 minutes)

- Let's revisit a problem that we examined in the last lesson.

Example 2

Water flows from a faucet into a bathtub at a constant rate of 7 gallons of water every 2 minutes Regard the volume of water accumulated in the tub as a function of the number of minutes the faucet has be on. Is this function discrete or not discrete?

- Assuming the tub is initially empty, we determined last lesson that the volume of water in the tub is given by $y = 3.5x$, where y is the volume of water in gallons, and x is the number of minutes the faucet has been on.
- What limitations are there on x and y?
    - *Both x and y should be positive numbers because they represent time and volume.*

This work is derived from Eureka Math ™ and licensed by Great Minds. ©2015 Great Minds. eureka-math.org
G8-M5-TE-B4-1.3.0-07.2015

- Would this function be considered discrete or not discrete? Explain.

  ▫ *This function is not discrete because we can assign any positive number to x, not just positive integers.*

Example 3 (8 minutes)

This is a more complicated example of a problem leading to a non-discrete function.

Scaffolding:

The more real you can make this, the better. Consider having a cooling cup of soup, coffee, or tea with a digital thermometer available for students to observe.

Example 3

You have just been served freshly made soup that is so hot that it cannot be eaten. You measure the temperature of the soup, and it is $210°$F. Since $212°$F is boiling, there is no way it can safely be eaten yet. One minute after receiving the soup, the temperature has dropped to $203°$F. If you assume that the rate at which the soup cools is constant, write an equation that would describe the temperature of the soup over time.

The temperature of the soup dropped $7°$F in one minute. Assuming the cooling continues at the same rate, then if y is the temperature of the soup after x minutes, then, $y = 210 - 7x$.

- We want to know how long it will be before the temperature of the soup is at a more tolerable temperature of $147°$F. The difference in temperature from $210°$F to $147°$F is $63°$F. For what number x will our function have the value 147?

  ▫ *$147 = 210 - 7x$; then $7x = 63$, and so $x = 9$.*

- Curious whether or not you are correct in assuming the cooling rate of the soup is constant, you decide to measure the temperature of the soup each minute after its arrival to you. Here's the data you obtain:

Time	Temperature in Fahrenheit
after 2 minutes	196
after 3 minutes	190
after 4 minutes	184
after 5 minutes	178
after 6 minutes	173
after 7 minutes	168
after 8 minutes	163
after 9 minutes	158

Our function led us to believe that after 9 minutes the soup would be safe to eat. The data in the table shows that it is still too hot.

- What do you notice about the change in temperature from one minute to the next?

  ▫ *For the first few minutes, minute 2 to minute 5, the temperature decreased $6°$F each minute. From minute 5 to minute 9, the temperature decreased just $5°$F each minute.*

- Since the rate of cooling at each minute is not constant, this function is said to be a nonlinear function.

- Sir Isaac Newton not only studied the motion of objects under gravity but also studied the rates of cooling of heated objects. He found that they do not cool at constant rates and that the functions that describe their temperature over time are indeed far from linear. (In fact, Newton's theory establishes that the temperature of soup at time x minutes would actually be given by the formula $y = 70 + 140 \left(\frac{133}{140}\right)^{x}$.)

Lesson 4:   More Examples of Functions

This work is derived from Eureka Math ™ and licensed by Great Minds. ©2015 Great Minds. eureka-math.org
G8-M5-TE-B4-1.3.0-07.2015

Example 4 (6 minutes)

Example 4

Consider the function that assigns to each of nine baseball players, numbered 1 through 9, his height. The data for this function is given below. Call the function G.

Player Number	Height
1	5′11″
2	5′4″
3	5′9″
4	5′6″
5	6′3″
6	6′8″
7	5′9″
8	5′10″
9	6′2″

- What output does G assign to the input 2?

  - *The function G assigns the height $5′4″$ to the player 2.*

- Could the function G simultaneously assign a second, different output to player 2? Explain.

  - *No. The function assigns height to a particular player. There is no way that a player can have two different heights.*

- It is not clear if there is a formula for this function. (And even if there were, it is not clear that it would be meaningful since who is labeled *player 1*, *player 2*, and so on is probably arbitrary.) In general, we can hope to have formulas for functions, but in reality we cannot expect to find them. (People would love to have a formula that explains and predicts the stock market, for example.)

- Can we classify this function as discrete or not discrete? Explain.

  - *This function would be described as discrete because the inputs are particular players.*

Exercises 1–3 (10 minutes)

Exercises 1–3

1. At a certain school, each bus in its fleet of buses can transport 35 students. Let B be the function that assigns to each count of students the number of buses needed to transport that many students on a field trip.

    When Jinpyo thought about matters, he drew the following table of values and wrote the formula $B = \frac{x}{35}$. Here x is the count of students, and B is the number of buses needed to transport that many students. He concluded that B is a linear function.

Number of students (x)	35	70	105	140
Number of buses (B)	1	2	3	4

This work is derived from Eureka Math ™ and licensed by Great Minds. ©2015 Great Minds. eureka-math.org
G8-M5-TE-B4-1.3.0-07.2015

Alicia looked at Jinpyo's work and saw no errors with his arithmetic. But she said that the function is not actually linear.

a. Alicia is right. Explain why B is not a linear function.

For 36 students, say, we'll need two buses—an extra bus for the extra student. In fact, for 36, 37, …, up to 70 students, the function B assigns the same value 2. For 71, 72, …, up to 105, it assigns the value 3. There is not a constant rate of increase of the buses needed, and so the function is not linear.

b. Is B a discrete function?

It is a discrete function.

2. A linear function has the table of values below. It gives the costs of purchasing certain numbers of movie tickets.

Number of tickets (x)	3	6	9	12
Total cost in dollars (y)	27.75	55.50	83.25	111.00

a. Write the linear function that represents the total cost, y, for x tickets purchased.

$$y = \frac{27.75}{3}x$$
$$y = 9.25x$$

b. Is the function discrete? Explain.

The function is discrete. You cannot have half of a movie ticket; therefore, it must be a whole number of tickets, which means it is discrete.

c. What number does the function assign to 4? What do the question and your answer mean?

It is asking us to determine the cost of buying 4 tickets. The function assigns 37 to 4. The answer means that 4 tickets will cost 37.00.

3. A function produces the following table of values.

Input	Output
Banana	B
Cat	C
Flippant	F
Oops	O
Slushy	S

a. Make a guess as to the rule this function follows. Each input is a word from the English language.

This function assigns to each word its first letter.

b. Is this function discrete?

It is discrete.

Lesson 4:   More Examples of Functions

EUREKA MATH

This work is derived from Eureka Math ™ and licensed by Great Minds. ©2015 Great Minds. eureka-math.org
G8-M5-TE-B4-1.3.0-07.2015

55

Closing (4 minutes)

Summarize, or ask students to summarize, the main points from the lesson:

- We have classified functions as either discrete or not discrete.
- Discrete functions admit only individually separate input values (such as whole numbers of students, or words of the English language). Functions that are not discrete admit any input value within a range of values (fractional values, for example).
- Functions that describe motion or smooth changes over time, for example, are typically not discrete.

> **Lesson Summary**
>
> **Functions are classified as either discrete or not discrete.**
>
> **Discrete functions admit only individually separate input values (such as whole numbers of students, or words of the English language). Functions that are not discrete admit any input value within a range of values (fractional values, for example).**
>
> **Functions that describe motion or smooth changes over time, for example, are typically not discrete.**

Exit Ticket (4 minutes)

This work is derived from Eureka Math ™ and licensed by Great Minds. ©2015 Great Minds. eureka-math.org
G8-M5-TE-B4-1.3.0-07.2015

Name ____   Date ___________________

Lesson 4: More Examples of Functions

Exit Ticket

1. The table below shows the costs of purchasing certain numbers of tablets. We can assume that the total cost is a linear function of the number of tablets purchased.

Number of tablets (x)	17	22	25
Total cost in dollars (y)	10,183.00	13,178.00	14,975.00

   a.   Write an equation that describes the total cost, y, as a linear function of the number, x, of tablets purchased.

   b.   Is the function discrete? Explain.

   c.   What number does the function assign to 7? Explain.

2. A function C assigns to each word in the English language the number of letters in that word. For example, C assigns the number 6 to the word *action*.

   a.   Give an example of an input to which C would assign the value 3.

   b.   Is C a discrete function? Explain.

Lesson 4:   More Examples of Functions

This work is derived from Eureka Math ™ and licensed by Great Minds. ©2015 Great Minds. eureka-math.org
G8-M5-TE-B4-1.3.0-07.2015

Exit Ticket Sample Solutions

1. The table below shows the costs of purchasing certain numbers of tablets. We can assume that the total cost is a linear function of the number of tablets purchased.

Number of tablets (x)	17	22	25
Total cost in dollars (y)	10,183.00	13,178.00	14,975.00

a. Write an equation that described the total cost, y, as a linear function of the number, x, of tablets purchased.

$$y = \frac{10,183}{17}x$$
$$y = 599x$$

b. Is the function discrete? Explain.

The function is discrete. You cannot have half of a tablet; therefore, it must be a whole number of tablets, which means it is discrete.

c. What number does the function assign to 7? Explain.

The function assigns $4,193$ to 7, which means that the cost of 7 tablets would be $\$4,193.00$.

2. A function C assigns to each word in the English language the number of letters in that word. For example, C assigns the number 6 to the word *action*.

a. Give an example of an input to which C would assign the value 3.

Any three-letter word will do.

b. Is C a discrete function? Explain.

The function is discrete. The input is a word in the English language, therefore it must be an entire word, not part of one, which means it is discrete.

This work is derived from Eureka Math ™ and licensed by Great Minds. ©2015 Great Minds. eureka-math.org
G8-M5-TE-B4-1.3.0-07.2015

Problem Set Sample Solutions

1. The costs of purchasing certain volumes of gasoline are shown below. We can assume that there is a linear relationship between x, the number of gallons purchased, and y, the cost of purchasing that many gallons.

Number of gallons (x)	5.4	6	15	17
Total cost in dollars (y)	19.71	21.90	54.75	62.05

   a. Write an equation that describes y as a linear function of x.

   $$y = 3.65x$$

   b. Are there any restrictions on the values x and y can adopt?

   Both x and y must be positive rational numbers.

   c. Is the function discrete?

   The function is not discrete.

   d. What number does the linear function assign to 20? Explain what your answer means.

   $$y = 3.65(20)$$
   $$y = 73$$

   The function assigns 73 to 20. It means that if 20 gallons of gas are purchased, it will cost $73.00.

2. A function has the table of values below. Examine the information in the table to answer the questions below.

Input	Output
one	3
two	3
three	5
four	4
five	4
six	3
seven	5

   a. Describe the function.

   The function assigns those particular numbers to those particular seven words. We don't know if the function accepts any more inputs and what it might assign to those additional inputs. (Though it does seem compelling to say that this function assigns to each positive whole number the count of letters in the name of that whole number.)

   b. What number would the function assign to the word *eleven*?

   We do not have enough information to tell. We are not even sure if eleven is considered a valid input for this function.

Lesson 4:   More Examples of Functions

EUREKA MATH™

This work is derived from Eureka Math ™ and licensed by Great Minds. ©2015 Great Minds. eureka-math.org
G8-M5-TE-B4-1.3.0-07.2015

3. The table shows the distances covered over certain counts of hours traveled by a driver driving a car at a constant speed.

Number of hours driven (x)	3	4	5	6
Total miles driven (y)	141	188	235	282

a. Write an equation that describes y, the number of miles covered, as a linear function of x, number of hours driven.

$$y = \frac{141}{3}x$$
$$y = 47x$$

b. Are there any restrictions on the value x and y can adopt?

Both x and y must be positive rational numbers.

c. Is the function discrete?

The function is not discrete.

d. What number does the function assign to 8? Explain what your answer means.

$$y = 47(8)$$
$$y = 376$$

The function assigns 376 to 8. The answer means that 376 miles are driven in 8 hours.

e. Use the function to determine how much time it would take to drive 500 miles.

$$500 = 47x$$
$$\frac{500}{47} = x$$
$$10.63829\ldots = x$$
$$10.6 \approx x$$

It would take about 10.6 hours to drive 500 miles.

EUREKA MATH™

This work is derived from Eureka Math ™ and licensed by Great Minds. ©2015 Great Minds. eureka-math.org
G8-M5-TE-B4-1.3.0-07.2015

4. Consider the function that assigns to each time of a particular day the air temperature at a specific location in Ithaca, NY. The following table shows the values of this function at some specific times.

12:00 noon	$92°F$
1:00 p.m.	$90.5°F$
2:00 p.m.	$89°F$
4:00 p.m.	$86°F$
8:00 p.m.	$80°F$

a. Let y represent the air temperature at time x hours past noon. Verify that the data in the table satisfies the linear equation $y = 92 - 1.5x$.

At 12:00, 0 hours have passed since 12:00; then, $y = 92 - 1.5(0) = 92$.

At 1:00, 1 hour has passed since 12:00; then, $y = 92 - 1.5(1) = 90.5$.

At 2:00, 2 hours have passed since 12:00; then, $y = 92 - 1.5(2) = 89$.

At 4:00, 4 hours have passed since 12:00; then, $y = 92 - 1.5(4) = 86$.

At 8:00, 8 hours have passed since 12:00; then, $y = 92 - 1.5(8) = 80$.

b. Are there any restrictions on the types of values x and y can adopt?

The input is a particular time of the day, and y is the temperature. The input cannot be negative but could be intervals that are fractions of an hour. The output could potentially be negative because it can get that cold.

c. Is the function discrete?

The function is not discrete.

d. According to the linear function of part (a), what will the air temperature be at 5:30 p.m.?

At 5:30, 5.5 hours have passed since 12:00; then $y = 92 - 1.5(5.5) = 83.75$.

The temperature at 5:30 will be $83.75°F$.

e. Is it reasonable to assume that this linear function could be used to predict the temperature for 10:00 a.m. the following day or a temperature at any time on a day next week? Give specific examples in your explanation.

No. There is no reason to expect this function to be linear. Temperature typically fluctuates and will, for certain, rise at some point.

We can show that our model for temperature is definitely wrong by looking at the predicted temperature one week (168 hours) later:

$$y = 92 - 1.5(168)$$

$$y = -160.$$

This is an absurd prediction.

Lesson 4:   More Examples of Functions

This work is derived from Eureka Math ™ and licensed by Great Minds. ©2015 Great Minds. eureka-math.org
G8-M5-TE-B4-1.3.0-07.2015

Lesson 5: Graphs of Functions and Equations

Student Outcomes

- Students define the graph of a numerical function to be the set of all points (x, y) with x an input of the function and y its matching output.
- Students realize that if a numerical function can be described by an equation, then the graph of the function precisely matches the graph of the equation.

Classwork

Exploratory Challenge/Exercises 1–3 (15 minutes)

Students work independently or in pairs to complete Exercises 1–3.

Exploratory Challenge/Exercises 1–3

1. The distance that Giselle can run is a function of the amount of time she spends running. Giselle runs 3 miles in 21 minutes. Assume she runs at a constant rate.

   a. Write an equation in two variables that represents her distance run, y, as a function of the time, x, she spends running.

   $$\frac{3}{21} = \frac{y}{x}$$
   $$y = \frac{1}{7}x$$

   b. Use the equation you wrote in part (a) to determine how many miles Giselle can run in 14 minutes.

   $$y = \frac{1}{7}(14)$$
   $$y = 2$$

   Giselle can run 2 miles in 14 minutes.

   c. Use the equation you wrote in part (a) to determine how many miles Giselle can run in 28 minutes.

   $$y = \frac{1}{7}(28)$$
   $$y = 4$$

   Giselle can run 4 miles in 28 minutes.

   d. Use the equation you wrote in part (a) to determine how many miles Giselle can run in 7 minutes.

   $$y = \frac{1}{7}(7)$$
   $$y = 1$$

   Giselle can run 1 mile in 7 minutes.

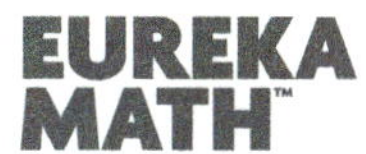

This work is derived from Eureka Math ™ and licensed by Great Minds. ©2015 Great Minds. eureka-math.org
G8-M5-TE-B4-1.3.0-07.2015

e. For a given input x of the function, a time, the matching output of the function, y, is the distance Giselle ran in that time. Write the inputs and outputs from parts (b)–(d) as ordered pairs, and plot them as points on a coordinate plane.

$$(14, 2), (28, 4), (7, 1)$$

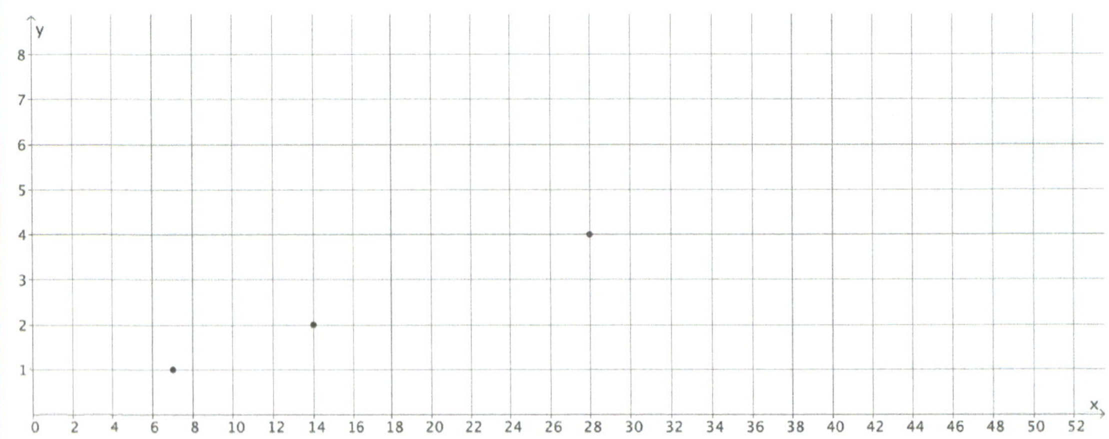

f. What do you notice about the points you plotted?

The points appear to be in a line.

g. Is the function discrete?

The function is not discrete because we can find the distance Giselle runs for any given amount of time she spends running.

h. Use the equation you wrote in part (a) to determine how many miles Giselle can run in 36 minutes. Write your answer as an ordered pair, as you did in part (e), and include the point on the graph. Is the point in a place where you expected it to be? Explain.

$$y = \frac{1}{7}(36)$$
$$y = \frac{36}{7}$$
$$y = 5\frac{1}{7}$$

$\left(36, 5\frac{1}{7}\right)$ *The point is where I expected it to be because it is in line with the other points.*

i. Assume you used the rule that describes the function to determine how many miles Giselle can run for any given time and wrote each answer as an ordered pair. Where do you think these points would appear on the graph?

I think all of the points would fall on a line.

j. What do you think the graph of all the possible input/output pairs would look like? Explain.

I know the graph will be a line as we can find all of the points that represent fractional intervals of time too. We also know that Giselle runs at a constant rate, so we would expect that as the time she spends running increases, the distance she can run will increase at the same rate.

EUREKA MATH

This work is derived from Eureka Math ™ and licensed by Great Minds. ©2015 Great Minds. eureka-math.org
G8-M5-TE-B4-1.3.0-07.2015

k. Connect the points you have graphed to make a line. Select a point on the graph that has integer coordinates. Verify that this point has an output that the function would assign to the input.

Answers will vary. Sample student work:

The point $(42, 6)$ is a point on the graph.

$$y = \frac{1}{7}x$$
$$6 = \frac{1}{7}(42)$$
$$6 = 6$$

The function assigns the output of 6 to the input of 42.

l. Sketch the graph of the equation $y = \frac{1}{7}x$ using the same coordinate plane in part (e). What do you notice about the graph of all the input/output pairs that describes Giselle's constant rate of running and the graph of the equation $y = \frac{1}{7}x$?

The graphs of the equation and the function coincide completely.

2. Sketch the graph of the equation $y = x^2$ for positive values of x. Organize your work using the table below, and then answer the questions that follow.

x	y
0	0
1	1
2	4
3	9
4	16
5	25
6	36

a. Plot the ordered pairs on the coordinate plane.

b. What shape does the graph of the points appear to take?

It appears to take the shape of a curve.

c. Is this equation a linear equation? Explain.

No, the equation $y = x^2$ is not a linear equation because the exponent of x is greater than 1.

d. Consider the function that assigns to each square of side length s units its area A square units. Write an equation that describes this function.

$$A = s^2$$

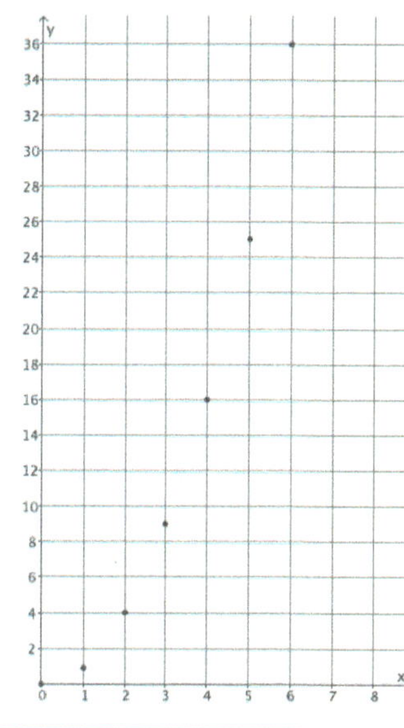

EUREKA MATH™

This work is derived from Eureka Math ™ and licensed by Great Minds. ©2015 Great Minds. eureka-math.org
G8-M5-TE-B4-1.3.0-07.2015

e.   What do you think the graph of all the input/output pairs (s, A) of this function will look like? Explain.

I think the graph of input/output pairs will look like the graph of the equation $y = x^2$. The inputs and outputs would match the solutions to the equation exactly. For the equation, the y value is the square of the x value. For the function, the output is the square of the input.

f.   Use the function you wrote in part (d) to determine the area of a square with side length 2.5 units. Write the input and output as an ordered pair. Does this point appear to belong to the graph of $y = x^2$?

$$A = (2.5)^2$$
$$A = 6.25$$

The area of the square is 6.25 units squared. $(2.5, 6.25)$ The point looks like it would belong to the graph of $y = x^2$; it looks like it would be on the curve that the shape of the graph is taking.

3.   The number of devices a particular manufacturing company can produce is a function of the number of hours spent making the devices. On average, 4 devices are produced each hour. Assume that devices are produced at a constant rate.

a.   Write an equation in two variables that describes the number of devices, y, as a function of the time the company spends making the devices, x.

$$\frac{4}{1} = \frac{y}{x}$$
$$y = 4x$$

b.   Use the equation you wrote in part (a) to determine how many devices are produced in 8 hours.

$$y = 4(8)$$
$$y = 32$$

The company produces 32 devices in 8 hours.

c.   Use the equation you wrote in part (a) to determine how many devices are produced in 6 hours.

$$y = 4(6)$$
$$y = 24$$

The company produces 24 devices in 6 hours.

d.   Use the equation you wrote in part (a) to determine how many devices are produced in 4 hours.

$$y = 4(4)$$
$$y = 16$$

The company produces 16 devices in 4 hours.

e.   The input of the function, x, is time, and the output of the function, y, is the number of devices produced. Write the inputs and outputs from parts (b)–(d) as ordered pairs, and plot them as points on a coordinate plane.

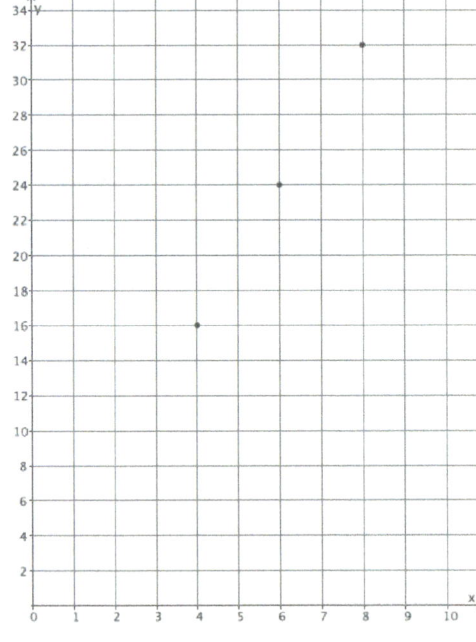

$(8, 32), (6, 24), (4, 16)$

Lesson 5:   Graphs of Functions and Equations

EUREKA MATH™

This work is derived from Eureka Math ™ and licensed by Great Minds. ©2015 Great Minds. eureka-math.org
G8-M5-TE-B4-1.3.0-07.2015

f.   **What shape does the graph of the points appear to take?**

The points appear to be in a line.

g.   **Is the function discrete?**

The function is not discrete because we can find the number of devices produced for any given time, including fractions of an hour.

h.   **Use the equation you wrote in part (a) to determine how many devices are produced in 1.5 hours. Write your answer as an ordered pair, as you did in part (e), and include the point on the graph. Is the point in a place where you expected it to be? Explain.**

$$y = 4(1.5)$$
$$y = 6$$

$(1.5, 6)$ *The point is where I expected it to be because it is in line with the other points.*

i.   **Assume you used the equation that describes the function to determine how many devices are produced for any given time and wrote each answer as an ordered pair. Where do you think these points would appear on the graph?**

I think all of the points would fall on a line.

j.   **What do you think the graph of all possible input/output pairs will look like? Explain.**

I think the graph of this function will be a line. Since the rate is continuous, we can find all of the points that represent fractional intervals of time. We also know that devices are produced at a constant rate, so we would expect that as the time spent producing devices increases, the number of devices produced would increase at the same rate.

k.   **Connect the points you have graphed to make a line. Select a point on the graph that has integer coordinates. Verify that this point has an output that the function would assign to the input.**

Answers will vary. Sample student work:

The point $(5, 20)$ is a point on the graph.

$$y = 4x$$
$$20 = 4(5)$$
$$20 = 20$$

The function assigns the output of 20 to the input of 5.

l.   **Sketch the graph of the equation $y = 4x$ using the same coordinate plane in part (e). What do you notice about the graph of input/output pairs that describes the company's constant rate of producing devices and the graph of the equation $y = 4x$?**

The graphs of the equation and the function coincide completely.

This work is derived from Eureka Math ™ and licensed by Great Minds. ©2015 Great Minds. eureka-math.org
G8-M5-TE-B4-1.3.0-07.2015

Discussion (10 minutes)

- What was the equation that described the function in Exercise 1, Giselle's distance run over given time intervals?

  □ *The equation was $y = \frac{1}{7}x$.*

- Given an input, how did you determine the output the function would assign?

  □ *We used the equation. In place of x, we put the input. The number that was computed was the output.*

MP.6

- So each input and its matching output correspond to a pair of numbers (x, y) that makes the equation $y = \frac{1}{7}x$ a true number sentence?

  □ *Yes*

Give students a moment to make sense of this, verifying that each pair of input/output values in Exercise 1 is indeed a pair of numbers (x, y) that make $y = \frac{1}{7}x$ a true statement.

- And suppose we have a pair of numbers (x, y) that make $y = \frac{1}{7}x$ a true statement with x positive. If x is an input of the function, the number of minutes Giselle runs, would y be its matching output, the distance she covers?

  □ *Yes. We computed the outputs precisely by following the equation $y = \frac{1}{7}x$. So y will be the matching output to x.*

- So can we conclude that any pair of numbers (x, y) that make the equation $y = \frac{1}{7}x$ a true number statement correspond to an input and its matching output for the function?

  □ *Yes*

- And, backward, any pair of numbers (x, y) that represent an input/output pair for the function is a pair of numbers that make the equation $y = \frac{1}{7}x$ a true number statement?

  □ *Yes*

- Can we make similar conclusions about Exercise 3, the function that gives the devices built over a given number of hours?

Give students time to verify that the conclusions about Exercise 3 are the same as the conclusions about Exercise 1. Then continue with the discussion.

- The function in Exercise 3 is described by the equation $y = 4x$.
- We have that the ordered pairs (x, y) that make the equation $y = 4x$ a true number sentence precisely match the ordered pairs (x, y) with x an input of the function and y its matching output.
- Recall, in previous work, we defined the *graph of an equation* to be the set of all ordered pairs (x, y) that make the equation a true number sentence. Today we define the *graph of a function* to be the set of all the ordered pairs (x, y) with x an input of the function and y its matching output.
- And our discussion today shows that if a function can be described by an equation, then the graph of the function is precisely the same as the graph of the equation.
- It is sometimes possible to draw the graph of a function even if there is no obvious equation describing the function. (Consider having students plot some points of the function that assigns to each positive whole number its first digit, for example.)

     Lesson 5:   Graphs of Functions and Equations

This work is derived from Eureka Math ™ and licensed by Great Minds. ©2015 Great Minds. eureka-math.org
G8-M5-TE-B4-1.3.0-07.2015

- For Exercise 2, you began by graphing the equation $y = x^2$ for positive values of x. What was the shape of the graph?

  □ *It looked curved.*

- The graph had a curve in it because it was not the graph of a linear equation. All linear equations graph as lines. That is what we learned in Module 4. Since this equation was not linear, we should expect it to graph as something other than a line.

- What did you notice about the ordered pairs of the equation $y = x^2$ and the inputs and corresponding outputs for the function $A = s^2$?

  □ *The ordered pairs were exactly the same for the equation and the function.*

- What does that mean about the graphs of functions, even those that are not linear?

  □ *It means that the graph of a function will be identical to the graph of an equation.*

Exploratory Challenge/Exercise 4 (7 minutes)

Students work in pairs to complete Exercise 4.

Exploratory Challenge/Exercise 4

4. Examine the three graphs below. Which, if any, could represent the graph of a function? Explain why or why not for each graph.

Graph 1:

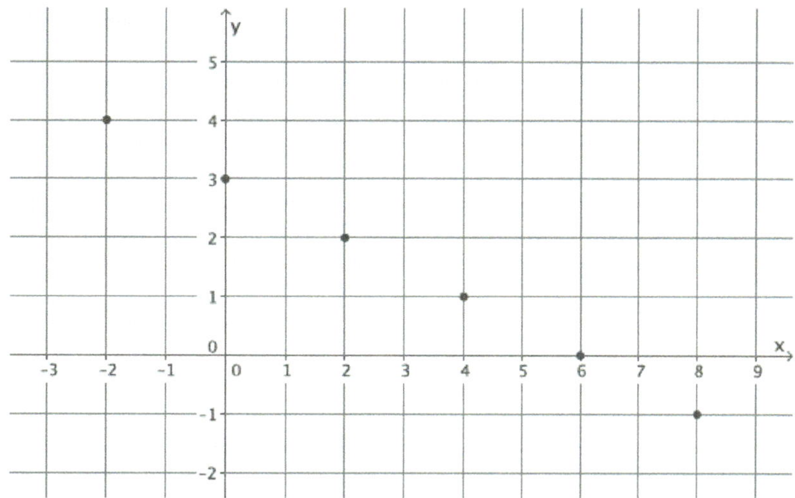

This is the graph of a function. Each input is a real number x, and we see from the graph that there is an output y to associate with each such input. For example, the ordered pair $(-2, 4)$ on the line associates the output 4 to the input -2.

This work is derived from Eureka Math ™ and licensed by Great Minds. ©2015 Great Minds. eureka-math.org
G8-M5-TE-B4-1.3.0-07.2015

Graph 2:

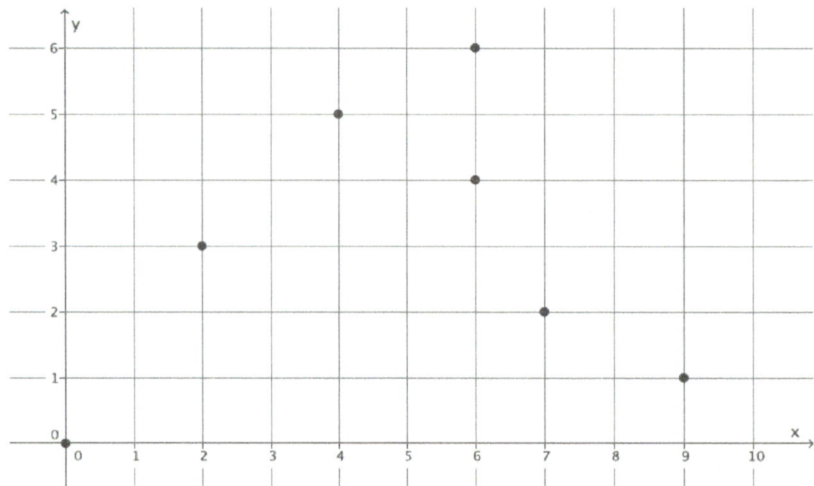

This is not the graph of a function. The ordered pairs $(6, 4)$ and $(6, 6)$ show that for the input of 6 there are two different outputs, both 4 and 6. We do not have a function.

Graph 3:

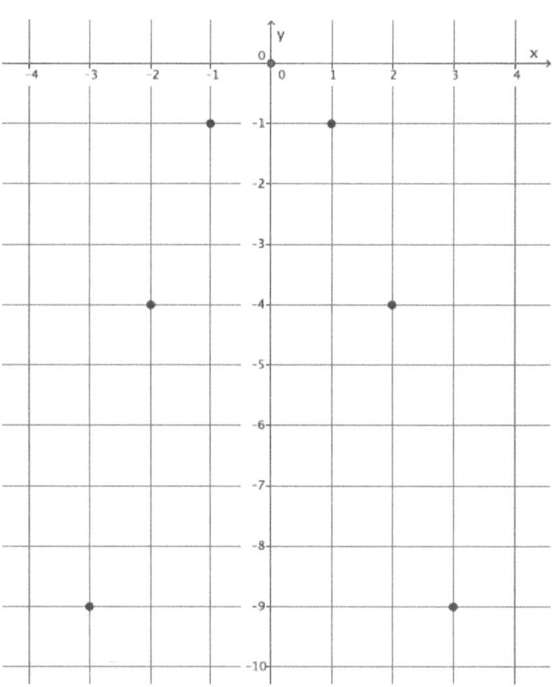

This is the graph of a function. The ordered pairs $(-3, -9)$, $(-2, -4)$, $(-1, -1)$, $(0, 0)$, $(1, -1)$, $(2, -4)$, and $(3, -9)$ represent inputs and their unique outputs.

Lesson 5:   Graphs of Functions and Equations

This work is derived from Eureka Math ™ and licensed by Great Minds. ©2015 Great Minds. eureka-math.org
G8-M5-TE-B4-1.3.0-07.2015

EUREKA MATH™

Discussion (3 minutes)

- The graph of a function is the set of all points (x, y) with x an input for the function and y its matching output. How did you use this definition to determine which graphs, if any, were functions?

    □ *By the definition of a function, we need each input to have only one output. On a graph, this means there cannot be two different ordered pairs with the same x value.*

- Assume the following set of ordered pairs is from some graph. Could this be the graph of a function? Explain.

$$(3, 5), (4, 7), (3, 9), (5, -2)$$

    □ *No, because the input of 3 has two different outputs. It does not fit the definition of a function.*

- Assume the following set of ordered pairs is from some graph. Could this be the graph of a function? Explain.

$$(-1, 6), (-3, 8), (5, 10), (7, 6)$$

    □ *Yes, it is possible as each input has a unique output. It satisfies the definition of a function so far.*

- Which of the following four graphs are functions? Explain.

Graph 1:

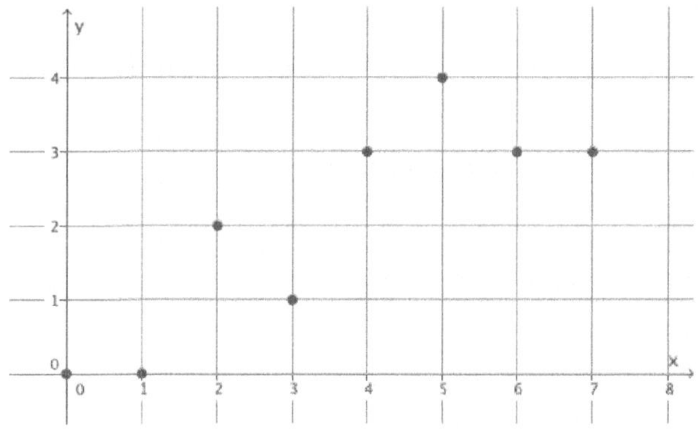

Graph 2:

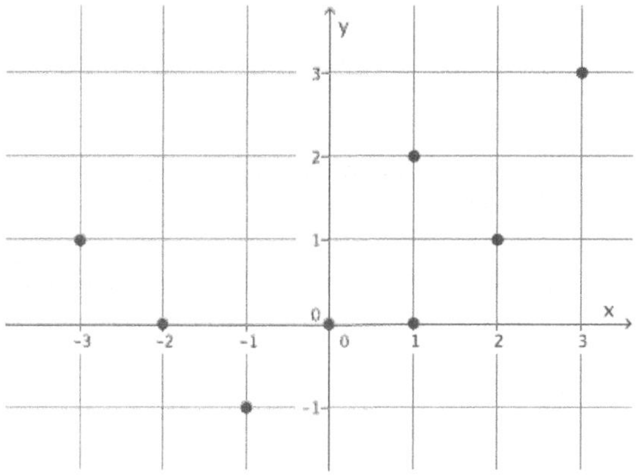

Graph 3:

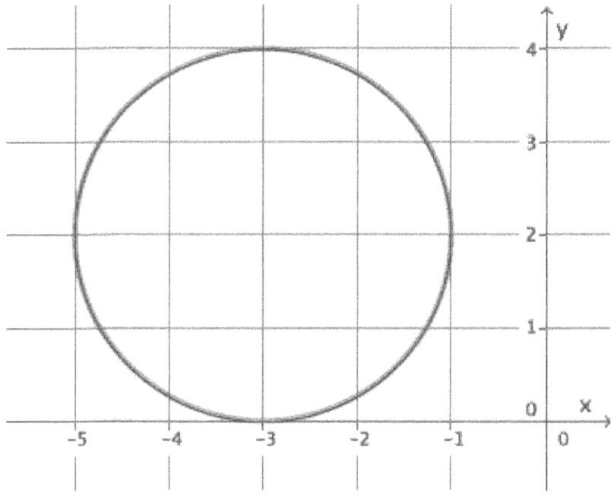

Graph 4:

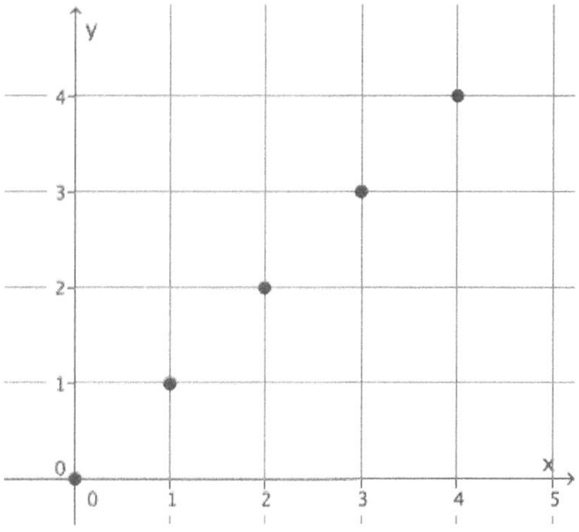

This work is derived from Eureka Math ™ and licensed by Great Minds. ©2015 Great Minds. eureka-math.org
G8-M5-TE-B4-1.3.0-07.2015

▫ *Graphs 1 and 4 are functions. Graphs 2 and 3 are not. Graphs 1 and 4 show that for each input of x, there is a unique output of y. For Graph 2, the input of $x = 1$ has two different outputs, $y = 0$ and $y = 2$, which means it cannot be a function. For Graph 3, it appears that each value of x between -5 and -1, excluding -5 and -1, has two outputs, one on the lower half of the circle and one on the upper half, which means it does not fit the definition of function.*

Closing (5 minutes)

Summarize, or ask students to summarize, the main points from the lesson:

- The graph of a function is defined to be the set of all points (x, y) with x an input for the function and y its matching output.
- If a function can be described by an equation, then the graph of the function matches the graph of the equation (at least at points which correspond to valid inputs of the function).
- We can look at plots of points and determine if they could be the graphs of functions.

> **Lesson Summary**
>
> **The graph of a function is defined to be the set of all points (x, y) with x an input for the function and y its matching output.**
>
> **If a function can be described by an equation, then the graph of the function is the same as the graph of the equation that represents it (at least at points which correspond to valid inputs of the function).**
>
> **It is not possible for two different points in the plot of the graph of a function to have the same x-coordinate.**

Exit Ticket (5 minutes)

      Lesson 5:   Graphs of Functions and Equations

This work is derived from Eureka Math ™ and licensed by Great Minds. ©2015 Great Minds. eureka-math.org
G8-M5-TE-B4-1.3.0-07.2015

Name ____   Date ____________________

Lesson 5: Graphs of Functions and Equations

Exit Ticket

Water flows from a hose at a constant rate of 11 gallons every 4 minutes. The total amount of water that flows from the hose is a function of the number of minutes you are observing the hose.

a. Write an equation in two variables that describes the amount of water, y, in gallons, that flows from the hose as a function of the number of minutes, x, you observe it.

b. Use the equation you wrote in part (a) to determine the amount of water that flows from the hose during an 8-minute period, a 4-minute period, and a 2-minute period.

c. An input of the function, x, is time in minutes, and the output of the function, y, is the amount of water that flows out of the hose in gallons. Write the inputs and outputs from part (b) as ordered pairs, and plot them as points on the coordinate plane.

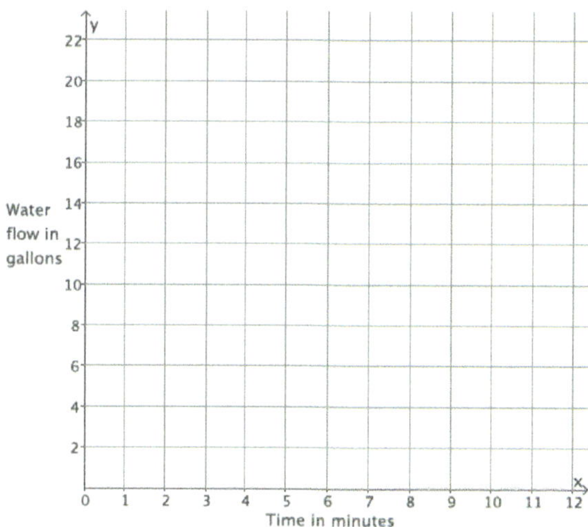

This work is derived from Eureka Math ™ and licensed by Great Minds. ©2015 Great Minds. eureka-math.org
G8-M5-TE-B4-1.3.0-07.2015

Exit Ticket Sample Solutions

Water flows from a hose at a constant rate of 11 gallons every 4 minutes. The total amount of water that flows from the hose is a function of the number of minutes you are observing the hose.

a. Write an equation in two variables that describes the amount of water, y, in gallons, that flows from the hose as a function of the number of minutes, x, you observe it.

$$\frac{11}{4} = \frac{y}{x}$$

$$y = \frac{11}{4}x$$

b. Use the equation you wrote in part (a) to determine the amount of water that flows from the hose during an 8-minute period, a 4-minute period, and a 2-minute period.

$$y = \frac{11}{4}(8)$$

$$y = 22$$

In 8 minutes, 22 gallons of water flow out of the hose.

$$y = \frac{11}{4}(4)$$

$$y = 11$$

In 4 minutes, 11 gallons of water flow out of the hose.

$$y = \frac{11}{4}(2)$$

$$y = 5.5$$

In 2 minutes, 5.5 gallons of water flow out of the hose.

c. An input of the function, x, is time in minutes, and the output of the function, y, is the amount of water that flows out of the hose in gallons. Write the inputs and outputs from part (b) as ordered pairs, and plot them as points on the coordinate plane.

$(8, 22), (4, 11), (2, 5.5)$

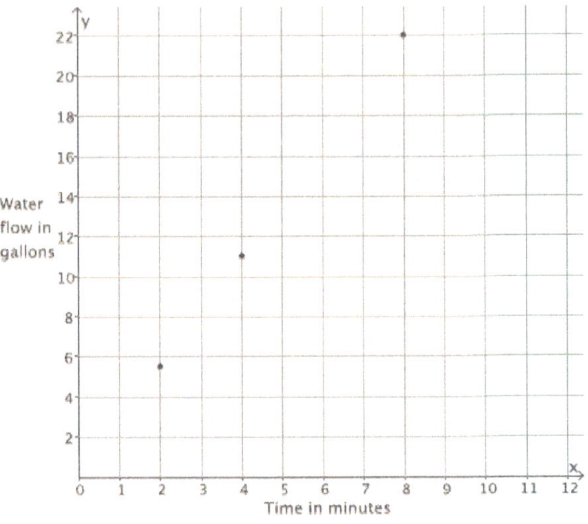

Lesson 5:   Graphs of Functions and Equations

EUREKA MATH

This work is derived from Eureka Math ™ and licensed by Great Minds. ©2015 Great Minds. eureka-math.org
G8-M5-TE-B4-1.3.0-07.2015

Problem Set Sample Solutions

1. The distance that Scott walks is a function of the time he spends walking. Scott can walk $\frac{1}{2}$ mile every 8 minutes. Assume he walks at a constant rate.

   a. Predict the shape of the graph of the function. Explain.

      The graph of the function will likely be a line because a linear equation can describe Scott's motion, and I know that the graph of the function will be the same as the graph of the equation.

   b. Write an equation to represent the distance that Scott can walk in miles, y, in x minutes.

      $$\frac{0.5}{8} = \frac{y}{x}$$
      $$y = \frac{0.5}{8}x$$
      $$y = \frac{1}{16}x$$

   c. Use the equation you wrote in part (b) to determine how many miles Scott can walk in 24 minutes.

      $$y = \frac{1}{16}(24)$$
      $$y = 1.5$$

      Scott can walk 1.5 miles in 24 minutes.

   d. Use the equation you wrote in part (b) to determine how many miles Scott can walk in 12 minutes.

      $$y = \frac{1}{16}(12)$$
      $$y = \frac{3}{4}$$

      Scott can walk 0.75 miles in 12 minutes.

   e. Use the equation you wrote in part (b) to determine how many miles Scott can walk in 16 minutes.

      $$y = \frac{1}{16}(16)$$
      $$y = 1$$

      Scott can walk 1 mile in 16 minutes.

This work is derived from Eureka Math ™ and licensed by Great Minds. ©2015 Great Minds. eureka-math.org
G8-M5-TE-B4-1.3.0-07.2015

f. Write your inputs and corresponding outputs as ordered pairs, and then plot them on a coordinate plane.

$(24, 1.5), (12, 0.75), (16, 1)$

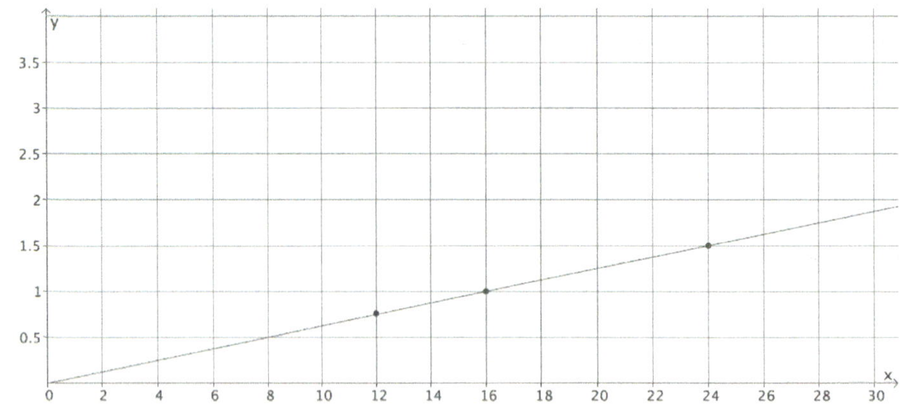

g. What shape does the graph of the points appear to take? Does it match your prediction?

The points appear to be in a line. Yes, as I predicted, the graph of the function is a line.

h. Connect the points to make a line. What is the equation of the line?

It is the equation that described the function: $y = \frac{1}{16}x.$

Lesson 5:   Graphs of Functions and Equations

This work is derived from Eureka Math ™ and licensed by Great Minds. ©2015 Great Minds. eureka-math.org
G8-M5-TE-B4-1.3.0-07.2015

EUREKA MATH

2. Graph the equation $y = x^3$ for positive values of x. Organize your work using the table below, and then answer the questions that follow.

x	y
0	0
0.5	0.125
1	1
1.5	3.375
2	8
2.5	15.625

a. Plot the ordered pairs on the coordinate plane.

b. What shape does the graph of the points appear to take?

   It appears to take the shape of a curve.

c. Is this the graph of a linear function? Explain.

   No, this is not the graph of a linear function. The equation $y = x^3$ is not a linear equation.

d. Consider the function that assigns to each positive real number s the volume V of a cube with side length s units. An equation that describes this function is $V = s^3$. What do you think the graph of this function will look like? Explain.

   I think the graph of this function will look like the graph of the equation $y = x^3$. The inputs and outputs would match the solutions to the equation exactly. For the equation, the y-value is the cube of the x-value. For the function, the output is the cube of the input.

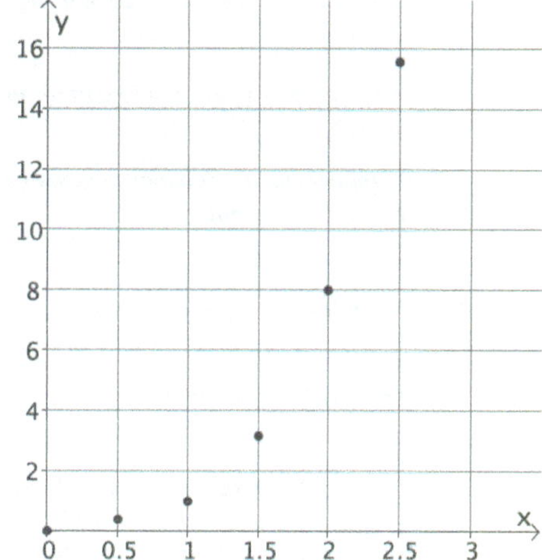

e. Use the function in part (d) to determine the volume of a cube with side length of 3 units. Write the input and output as an ordered pair. Does this point appear to belong to the graph of $y = x^3$?

$$V = (3)^3$$
$$V = 27$$

   $(3, 27)$ *The point looks like it would belong to the graph of $y = x^3$; it looks like it would be on the curve that the shape of the graph is taking.*

This work is derived from Eureka Math ™ and licensed by Great Minds. ©2015 Great Minds. eureka-math.org
G8-M5-TE-B4-1.3.0-07.2015

3. Sketch the graph of the equation $y = 180(x - 2)$ for whole numbers. Organize your work using the table below, and then answer the questions that follow.

x	y
3	180
4	360
5	540
6	720

a. Plot the ordered pairs on the coordinate plane.

b. What shape does the graph of the points appear to take?

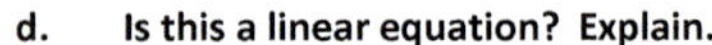

It appears to take the shape of a line.

c. Is this graph a graph of a function? How do you know?

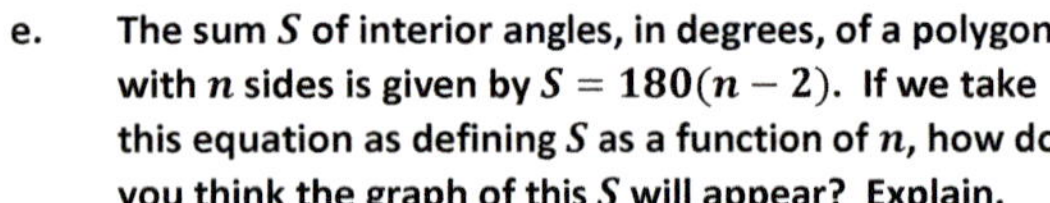

It appears to be a function because each input has exactly one output.

d. Is this a linear equation? Explain.

Yes, $y = 180(x - 2)$ is a linear equation. It can be rewritten as $y = 180x - 360$.

e. The sum S of interior angles, in degrees, of a polygon with n sides is given by $S = 180(n - 2)$. If we take this equation as defining S as a function of n, how do you think the graph of this S will appear? Explain.

I think the graph of this function will look like the graph of the equation $y = 180(x - 2)$. The inputs and outputs would match the solutions to the equation exactly.

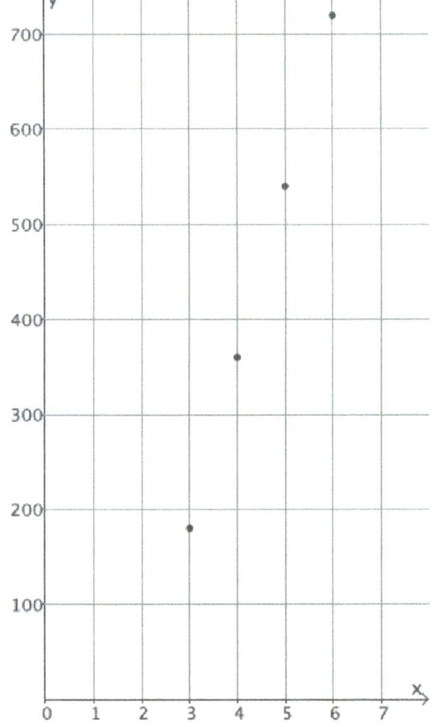

f. Is this function discrete? Explain.

The function $S = 180(n - 2)$ is discrete. The inputs are the number of sides, which are integers. The input, n, must be greater than 2 since three sides is the smallest number of sides for a polygon.

Lesson 5:   Graphs of Functions and Equations

EUREKA MATH™

This work is derived from Eureka Math ™ and licensed by Great Minds. ©2015 Great Minds. eureka-math.org
G8-M5-TE-B4-1.3.0-07.2015

4. Examine the graph below. Could the graph represent the graph of a function? Explain why or why not.

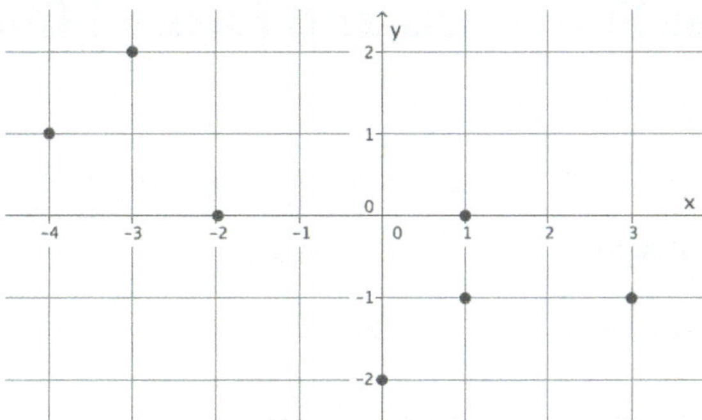

This is not the graph of a function. The ordered pairs $(1, 0)$ and $(1, -1)$ show that for the input of 1 there are two different outputs, both 0 and -1. For that reason, this cannot be the graph of a function because it does not fit the definition of a function.

5. Examine the graph below. Could the graph represent the graph of a function? Explain why or why not.

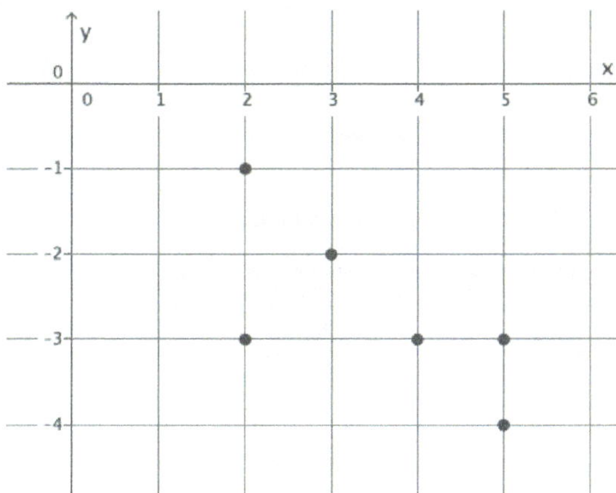

This is not the graph of a function. The ordered pairs $(2, -1)$ and $(2, -3)$ show that for the input of 2 there are two different outputs, both -1 and -3. Further, the ordered pairs $(5, -3)$ and $(5, -4)$ show that for the input of 5 there are two different outputs, both -3 and -4. For these reasons, this cannot be the graph of a function because it does not fit the definition of a function.

6. Examine the graph below. Could the graph represent the graph of a function? Explain why or why not.

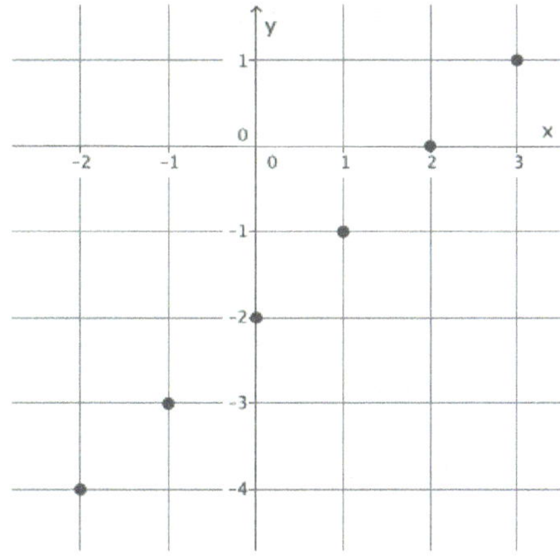

This is the graph of a function. The ordered pairs $(-2, -4)$, $(-1, -3)$, $(0, -2)$, $(1, -1)$, $(2, 0)$, and $(3, 1)$ represent inputs and their unique outputs. By definition, this is a function.

This work is derived from Eureka Math ™ and licensed by Great Minds. ©2015 Great Minds. eureka-math.org
G8-M5-TE-B4-1.3.0-07.2015

Lesson 6: Graphs of Linear Functions and Rate of Change

Student Outcomes

- Students deepen their understanding of linear functions.

Lesson Notes

This lesson contains a ten-minute fluency exercise that can occur at any time throughout this lesson. The exercise has students look for and make use of structure while solving multi-step equations.

Classwork

Opening Exercise (5 minutes)

> **Opening Exercise**
>
> A function is said to be linear if the rule defining the function can be described by a linear equation.
>
> Functions 1, 2, and 3 have table-values as shown. Which of these functions appear to be linear? Justify your answers.
>
> | Input | Output |
> |-------|--------|
> | 2 | 5 |
> | 4 | 7 |
> | 5 | 8 |
> | 8 | 11 |
>
> | Input | Output |
> |-------|--------|
> | 2 | 4 |
> | 3 | 9 |
> | 4 | 16 |
> | 5 | 25 |
>
> | Input | Output |
> |-------|--------|
> | 0 | −3 |
> | 1 | 1 |
> | 2 | 6 |
> | 3 | 9 |

Lead a short discussion that allows students to share their conjectures and reasoning. Revisit the Opening Exercise at the end of the discussion so students can verify if their conjectures were correct. Only the first function is a linear function.

Discussion (15 minutes)

Ask students to summarize what they learned from the last lesson. Make sure they recall that the graph of a numerical function is the set of ordered pairs (x, y) with x an input of the function and y its corresponding output. Also, recall that the graph of a function is identical to the graph of the equation that describes it (if there is one). Next, ask students to recall what they know about rate of change and slope and to recall that the graph of a linear equation (the set of pairs (x, y) that make the equation a true number sentence) is a straight line.

> *Scaffolding:*
>
> Students may need a brief review of the terms related to linear equations.

Lesson 6:   Graphs of Linear Functions and Rate of Change

EUREKA MATH

This work is derived from Eureka Math ™ and licensed by Great Minds. ©2015 Great Minds. eureka-math.org
G8-M5-TE-B4-1.3.0-07.2015

- Suppose a function can be described by an equation in the form of $y = mx + b$ and assigns the values shown in the table below:

Input	Output
2	5
3.5	8
4	9
4.5	10

- The graph of a linear equation is a line, and so the graph of this function will be a line. How do we compute the slope of the graph of a line?

  - *To compute slope, we find the difference in y-values compared to the difference in x-values. We use the following formula:*

  $$m = \frac{y_1 - y_2}{x_1 - x_2}.$$

- And what is the slope of the line associated with this data? Using the first two rows of the table we get:

$$\frac{5 - 8}{2 - 3.5} = \frac{-3}{-1.5}$$
$$= 2$$

- To check, calculate the rate of change between rows two and three and rows three and four as well.

  - *Sample student work:*

$$\frac{8 - 9}{3.5 - 4} = \frac{-1}{-0.5} \qquad \text{or} \qquad \frac{9 - 10}{4 - 4.5} = \frac{-1}{-0.5}$$
$$= 2 \qquad\qquad\qquad\qquad = 2$$

- Does the claim that the function is linear seem reasonable?

  - *Yes, the rate of change between each pair of inputs and outputs does seem to be constant.*

- Check one more time by computing the slope from one more pair.

  - *Sample student work:*

$$\frac{5 - 10}{2 - 4.5} = \frac{-5}{-2.5} \qquad \text{or} \qquad \frac{5 - 9}{2 - 4} = \frac{-4}{-2}$$
$$= 2 \qquad\qquad\qquad\qquad = 2$$

- Can we now find the equation that describes the function? At this point, we expect the equation to be described by $y = 2x + b$ because we know the slope is 2. Since the function assigns 5 to 2, 8 to 3.5, and so on, we can use that information to determine the value of b.

Using the assignment of 5 to 2 we see:

$$5 = 2(2) + b$$
$$5 = 4 + b$$
$$1 = b$$

This work is derived from Eureka Math ™ and licensed by Great Minds. ©2015 Great Minds. eureka-math.org
G8-M5-TE-B4-1.3.0-07.2015

- Now that we know that $b = 1$, we can substitute into $y = 2x + b$, which results in the equation $y = 2x + 1$. The equation that describes the function is $y = 2x + 1$, and the function is a linear function. What would the graph of this function look like?

  - *It would be a line because the rule that describes the function in the form of $y = mx + b$ is an equation known to graph as a line.*

- The following table represents the outputs that a function would assign to given inputs. We want to know if the function is a linear function and, if so, what linear equation describes the function.

Input	Output
−2	4
3	9
4.5	20.25
5	25

- How should we begin? How do we check if the function is linear?

  - *We need to inspect the rate of change between pairs of inputs and their corresponding outputs and see if that value is constant.*

- Compare at least three pairs of inputs and their corresponding outputs.

  - *Sample student work:*

$$\frac{4 - 9}{-2 - 3} = \frac{-5}{-5} \qquad \frac{4 - 25}{-2 - 5} = \frac{-21}{-7} \qquad \frac{9 - 25}{3 - 5} = \frac{-16}{-2}$$

$$= 1 \qquad\qquad\qquad = 3 \qquad\qquad\qquad = 8$$

- What do you notice about the rate of change, and what does this mean about the function?

  - *The rate of change was different for each pair of inputs and outputs inspected, which means that it is not a linear function.*

- If this were a linear function, what would we expect to see?

  - *If this were a linear function, each inspection of the rate of change would result in the same number (similar to what we saw in the last problem, in which each result was 2).*

- We have enough evidence to conclude that this function is not a linear function. Would the graph of this function be a line? Explain.

  - *No, the graph of this function would not be a line. Only linear functions, whose equations are in the form of $y = mx + b$, graph as lines. Since this function does not have a constant rate of change, it will not graph as a line.*

Exercise (5 minutes)

Students work independently or in pairs to complete the exercise. Make sure to discuss with the class the subtle points made in parts (c) and (d) and their solutions.

Lesson 6:   Graphs of Linear Functions and Rate of Change

EUREKA MATH™

This work is derived from Eureka Math ™ and licensed by Great Minds. ©2015 Great Minds. eureka-math.org
G8-M5-TE-B4-1.3.0-07.2015

Exercise

A function assigns to the inputs shown the corresponding outputs given in the table below.

Input	Output
1	2
2	−1
4	−7
6	−13

a. Do you suspect the function is linear? Compute the rate of change of this data for at least three pairs of inputs and their corresponding outputs.

$$\frac{2-(-1)}{1-2} = \frac{3}{-1} \qquad \frac{-7-(-13)}{4-6} = \frac{6}{-2} \qquad \frac{2-(-7)}{1-4} = \frac{9}{-3}$$
$$= -3 \qquad\qquad\qquad = -3 \qquad\qquad\qquad = -3$$

Yes, the rate of change is the same when I check pairs of inputs and corresponding outputs. Each time it is equal to -3. *Since the rate of change is the same, then I know it is a linear function.*

b. What equation seems to describe the function?

Using the assignment of 2 *to* 1*:*

$$2 = -3(1) + b$$
$$2 = -3 + b$$
$$5 = b$$

The equation that seems to describe the function is $y = -3x + 5$.

c. As you did not verify that the rate of change is constant across <u>all</u> input/output pairs, check that the equation you found in part (a) does indeed produce the correct output for each of the four inputs 1, 2, 4, and 6.

For $x = 1$ *we have* $y = -3(1) + 5 = 2$.

For $x = 2$ *we have* $y = -3(2) + 5 = -1$.

For $x = 4$ *we have* $y = -3(4) + 5 = -7$.

For $x = 6$ *we have* $y = -3(6) + 5 = -13$.

These are correct.

d. What will the graph of the function look like? Explain.

The graph of the function will be a plot of four points lying on a common line. As we were not told about any other inputs for this function, we must assume for now that there are only these four input values for the function.

The four points lie on the line with equation $y = -3x + 5$.

This work is derived from Eureka Math ™ and licensed by Great Minds. ©2015 Great Minds. eureka-math.org
G8-M5-TE-B4-1.3.0-07.2015

Closing (5 minutes)

Summarize, or ask students to summarize, the main points from the lesson:

- We know that if the rate of change for pairs of inputs and corresponding outputs for a function is the same for all pairs, then the function is a linear function.
- We know that a linear function can be described by a linear equation $y = mx + b$.
- We know that the graph of a linear function will be a set of points all lying on a common line. If the linear function is discrete, then its graph will be a set of distinct collinear points. If the linear function is not discrete, then its graph will be a full straight line or a portion of the line (as appropriate for the context of the problem).

Lesson Summary

If the rate of change for pairs of inputs and corresponding outputs for a function is the same for all pairs (constant), then the function is a linear function. It can thus be described by a linear equation $y = mx + b$.

The graph of a linear function will be a set of points contained in a line. If the linear function is discrete, then its graph will be a set of distinct collinear points. If the linear function is not discrete, then its graph will be a full straight line or a portion of the line (as appropriate for the context of the problem).

Exit Ticket (5 minutes)

Fluency Exercise (10 minutes): Multi-Step Equations I

Rapid White Board Exchange (*RWBE):* In this exercise, students solve three sets of similar multi-step equations. Display the equations one at a time. Each equation should be solved in less than one minute; however, students may need slightly more time for the first set and less time for the next two sets if they notice the pattern. Consider having students work on personal white boards, and have them show their solutions for each problem. The three sets of equations and their answers are located at the end of the lesson. Refer to the Rapid White Board Exchanges section in the Module Overview for directions to administer a RWBE.

Lesson 6:   Graphs of Linear Functions and Rate of Change

This work is derived from Eureka Math ™ and licensed by Great Minds. ©2015 Great Minds. eureka-math.org
G8-M5-TE-B4-1.3.0-07.2015

Name ____   Date _________________

Lesson 6: Graphs of Linear Functions and Rate of Change

Exit Ticket

1. Sylvie claims that a function with the table of inputs and outputs below is a linear function. Is she correct? Explain.

Input	Output
-3	-25
2	10
5	31
8	54

2. A function assigns the inputs and corresponding outputs shown in the table to the right.

    a. Does the function appear to be linear? Check at least three pairs of inputs and their corresponding outputs.

Input	Output
-2	3
8	-2
10	-3
20	-8

This work is derived from Eureka Math ™ and licensed by Great Minds. ©2015 Great Minds. eureka-math.org
G8-M5-TE-B4-1.3.0-07.2015

b. Can you write a linear equation that describes the function?

c. What will the graph of the function look like? Explain.

Lesson 6:   Graphs of Linear Functions and Rate of Change

This work is derived from Eureka Math ™ and licensed by Great Minds. ©2015 Great Minds. eureka-math.org
G8-M5-TE-B4-1.3.0-07.2015

EUREKA
MATH

Exit Ticket Sample Solutions

1. Sylvie claims that the function with the table of inputs and outputs is a linear function. Is she correct? Explain.

Input	Output
−3	−25
2	10
5	31
8	54

$$\frac{-25-(10)}{-3-2}=\frac{-35}{-5} \qquad \frac{10-31}{2-5}=\frac{-21}{-3} \qquad \frac{31-54}{5-8}=\frac{-23}{-3}$$
$$=7 \qquad\qquad =7 \qquad\qquad =\frac{23}{3}$$

No, this is not a linear function. The rate of change was not the same for each pair of inputs and outputs inspected, which means that it is not a linear function.

2. A function assigns the inputs and corresponding outputs shown in the table below.

   a. Does the function appear to be linear? Check at least three pairs of inputs and their corresponding outputs.

Input	Output
−2	3
8	−2
10	−3
20	−8

$$\frac{3-(-2)}{-2-8}=\frac{5}{-10}=-\frac{1}{2}$$
$$\frac{-2-(-3)}{8-10}=\frac{1}{-2}=-\frac{1}{2}$$
$$\frac{-3-(-8)}{10-20}=\frac{5}{-10}=-\frac{1}{2}$$

Yes. The rate of change is the same when I check pairs of inputs and corresponding outputs. Each time it is equal to $-\frac{1}{2}$. Since the rate of change is the same for at least these three examples, the function could well be linear.

   b. Can you write a linear equation that describes the function?

   We suspect we have an equation of the form $y = -\frac{1}{2}x + b$. Using the assignment of 3 to −2:

$$3 = -\frac{1}{2}(-2) + b$$
$$3 = 1 + b$$
$$2 = b$$

   The equation that describes the function might be $y = -\frac{1}{2}x + 2$.

   Checking: When $x = -2$, we get $y = -\frac{1}{2}(-2) + 2 = 3$. When $x = 8$, we get $y = -\frac{1}{2}(8) + 2 = -2$. When $x = 10$, we get $y = -\frac{1}{2}(10) + 2 = -3$. When $x = 20$, we get $y = -\frac{1}{2}(20) + 2 = -8$.

   It works.

   c. What will the graph of the function look like? Explain.

   The graph of the function will be four distinct points all lying in a line. (They all lie on the line with equation $y = -\frac{1}{2}x + 2$).

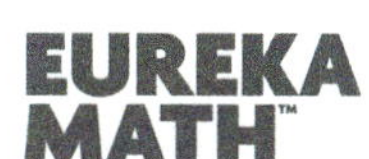

This work is derived from Eureka Math ™ and licensed by Great Minds. ©2015 Great Minds. eureka-math.org
G8-M5-TE-B4-1.3.0-07.2015

Problem Set Sample Solutions

1. A function assigns to the inputs given the corresponding outputs shown in the table below.

Input	Output
3	9
9	17
12	21
15	25

a. Does the function appear to be linear? Check at least three pairs of inputs and their corresponding outputs.

$$\frac{9-17}{3-9} = \frac{-8}{-6} \qquad\qquad \frac{17-21}{9-12} = \frac{-4}{-3} \qquad\qquad \frac{21-25}{12-15} = \frac{-4}{-3}$$

$$= \frac{4}{3} \qquad\qquad\qquad\qquad = \frac{4}{3} \qquad\qquad\qquad\qquad = \frac{4}{3}$$

Yes. The rate of change is the same when I check pairs of inputs and corresponding outputs. Each time it is equal to $\frac{4}{3}$. Since the rate of change is the same, the function does appear to be linear.

b. Find a linear equation that describes the function.

Using the assignment of 9 to 3

$$9 = \frac{4}{3}(3) + b$$
$$9 = 4 + b$$
$$5 = b$$

The equation that describes the function is $y = \frac{4}{3}x + 5$. (We check that for each of the four inputs given, this equation does indeed produce the correct matching output.)

c. What will the graph of the function look like? Explain.

The graph of the function will be four points in a row. They all lie on the line given by the equation $y = \frac{4}{3}x + 5$.

Lesson 6: Graphs of Linear Functions and Rate of Change

EUREKA MATH

This work is derived from Eureka Math ™ and licensed by Great Minds. ©2015 Great Minds. eureka-math.org
G8-M5-TE-B4-1.3.0-07.2015

2. A function assigns to the inputs given the corresponding outputs shown in the table below.

Input	Output
-1	2
0	0
1	2
2	8
3	18

a. Is the function a linear function?

$$\frac{2-0}{-1-0} = \frac{2}{-1} \qquad\qquad \frac{0-2}{0-1} = \frac{-2}{-1}$$
$$= -2 \qquad\qquad\qquad\qquad = 2$$

No. The rate of change is not the same when I check the first two pairs of inputs and corresponding outputs. All rates of change must be the same for all inputs and outputs for the function to be linear.

b. What equation describes the function?

I am not sure what equation describes the function. It is not a linear function.

3. A function assigns the inputs and corresponding outputs shown in the table below.

Input	Output
0.2	2
0.6	6
1.5	15
2.1	21

a. Does the function appear to be linear? Check at least three pairs of inputs and their corresponding outputs..

$$\frac{2-6}{0.2-0.6} = \frac{-4}{-0.4} \qquad \frac{6-15}{0.6-1.5} = \frac{-9}{-0.9} \qquad \frac{15-21}{1.5-2.1} = \frac{-6}{-0.6}$$
$$= 10 \qquad\qquad\qquad = 10 \qquad\qquad\qquad = 10$$

Yes. The rate of change is the same when I check pairs of inputs and corresponding outputs. Each time it is equal to 10. The function appears to be linear.

b. Find a linear equation that describes the function.

Using the assignment of 2 to 0.2:

$$2 = 10(0.2) + b$$
$$2 = 2 + b$$
$$0 = b$$

The equation that describes the function is $y = 10x$. It clearly fits the data presented in the table.

c. What will the graph of the function look like? Explain.

The graph will be four distinct points in a row. They all sit on the line given by the equation $y = 10x$.

This work is derived from Eureka Math ™ and licensed by Great Minds. ©2015 Great Minds. eureka-math.org
G8-M5-TE-B4-1.3.0-07.2015

4. Martin says that you only need to check the first and last input and output values to determine if the function is linear. Is he correct? Explain.

No, he is not correct. For example, consider the function with input and output values in this table.

Using the first and last input and output, the rate of change is

$$\frac{9 - 12}{1 - 3} = \frac{-3}{-2}$$
$$= \frac{3}{2}$$

Input	Output
1	9
2	10
3	12

But when you use the first two inputs and outputs, the rate of change is

$$\frac{9 - 10}{1 - 2} = \frac{-1}{-1}$$
$$= 1$$

Note to teacher: Accept any example where the rate of change is different for any two inputs and outputs.

5. Is the following graph a graph of a linear function? How would you determine if it is a linear function?

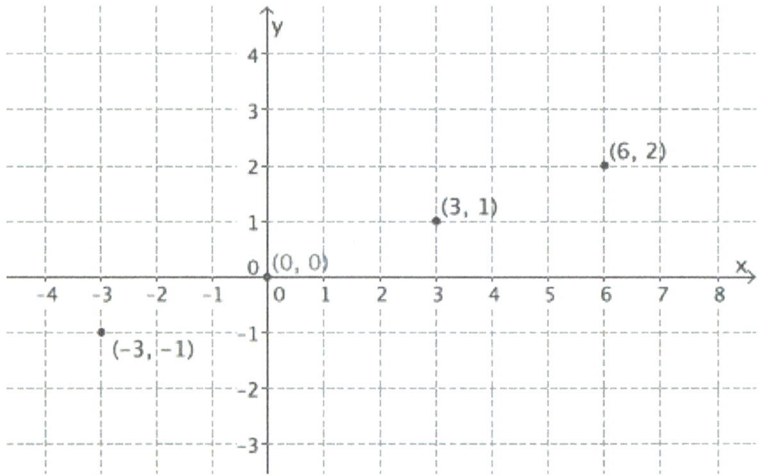

It appears to be a linear function. To check, I would organize the coordinates in an input and output table. Next, I would check to see that all the rates of change are the same. If they are the same rates of change, I would use the equation $y = mx + b$ and one of the assignments to write an equation to solve for b. That information would allow me to determine the equation that represents the function.

EUREKA MATH

This work is derived from Eureka Math ™ and licensed by Great Minds. ©2015 Great Minds. eureka-math.org
G8-M5-TE-B4-1.3.0-07.2015

6. A function assigns to the inputs given the corresponding outputs shown in the table below.

Input	Output
-6	-6
-5	-5
-4	-4
-2	-2

a. Does the function appear to be a linear function?

$$\frac{-6-(-5)}{-6-(-5)}=\frac{1}{1} \qquad \frac{-5-(-4)}{-5-(-4)}=\frac{1}{1} \qquad \frac{-4-(-2)}{-4-(-2)}=\frac{2}{2}$$

$$= 1 \qquad\qquad\qquad = 1 \qquad\qquad\qquad = 1$$

Yes. The rate of change is the same when I check pairs of inputs and corresponding outputs. Each time it is equal to 1. Since the rate of change is constant so far, it could be a linear function.

b. What equation describes the function?

Clearly the equation $y = x$ fits the data. It is a linear function.

c. What will the graph of the function look like? Explain.

The graph of the function will be four distinct points in a row. These four points lie on the line given by the equation $y = x$.

This work is derived from Eureka Math ™ and licensed by Great Minds. ©2015 Great Minds. eureka-math.org
G8-M5-TE-B4-1.3.0-07.2015

Multi-Step Equations I

Set 1:

$3x + 2 = 5x + 6$

$4(5x + 6) = 4(3x + 2)$

$$\frac{3x + 2}{6} = \frac{5x + 6}{6}$$

Answer for each problem in this set is $x = -2$.

Set 2:

$6 - 4x = 10x + 9$

$-2(-4x + 6) = -2(10x + 9)$

$$\frac{10x + 9}{5} = \frac{6 - 4x}{5}$$

Answer for each problem in this set is $x = -\dfrac{3}{14}$.

Set 3:

$5x + 2 = 9x - 18$

$8x + 2 - 3x = 7x - 18 + 2x$

$$\frac{2 + 5x}{3} = \frac{7x - 18 + 2x}{3}$$

Answer for each problem in this set is $x = 5$.

Lesson 6: Graphs of Linear Functions and Rate of Change

EUREKA MATH™

This work is derived from Eureka Math ™ and licensed by Great Minds. ©2015 Great Minds. eureka-math.org
G8-M5-TE-B4-1.3.0-07.2015

Lesson 7: Comparing Linear Functions and Graphs

Student Outcomes

- Students compare the properties of two functions that are represented in different ways via tables, graphs, equations, or written descriptions.
- Students use rate of change to compare linear functions.

Lesson Notes

The Fluency Exercise included in this lesson takes approximately 10 minutes and should be assigned either at the beginning or at the end of the lesson.

Classwork

Exploratory Challenge/Exercises 1–4 (20 minutes)

MP.1

Students work in small groups to complete Exercises 1–4. Groups can select a method of their choice to answer the questions and their methods will be a topic of discussion once the Exploratory Challenge is completed. Encourage students to discuss the various methods (e.g., graphing, comparing rates of change, using algebra) as a group before they begin solving.

Exploratory Challenge/Exercises 1–4

Each of Exercises 1–4 provides information about two functions. Use that information given to help you compare the two functions and answer the questions about them.

1. Alan and Margot each drive from City A to City B, a distance of 147 miles. They take the same route and drive at constant speeds. Alan begins driving at 1:40 p.m. and arrives at City B at 4:15 p.m. Margot's trip from City A to City B can be described with the equation $y = 64x$, where y is the distance traveled in miles and x is the time in minutes spent traveling. Who gets from City A to City B faster?

    Student solutions will vary. Sample solution is provided.

    It takes Alan 155 minutes to travel the 147 miles. Therefore, his constant rate is $\frac{147}{155}$ miles per minute.

    Margot drives 64 miles per hour (60 minutes). Therefore, her constant rate is $\frac{64}{60}$ miles per minute.

    To determine who gets from City A to City B faster, we just need to compare their rates in miles per minute.

    $$\frac{147}{155} < \frac{64}{60}$$

    Since Margot's rate is faster, she will get to City B faster than Alan.

> **Scaffolding:**
>
> Providing example language for students to reference will be useful. This might consist of sentence starters, sentence frames, or a word wall.

This work is derived from Eureka Math ™ and licensed by Great Minds. ©2015 Great Minds. eureka-math.org
G8-M5-TE-B4-1.3.0-07.2015

2. You have recently begun researching phone billing plans. Phone Company A charges a flat rate of $75 a month. A flat rate means that your bill will be $75 each month with no additional costs. The billing plan for Phone Company B is a linear function of the number of texts that you send that month. That is, the total cost of the bill changes each month depending on how many texts you send. The table below represents some inputs and the corresponding outputs that the function assigns.

Input (number of texts)	Output (cost of bill in dollars)
50	50
150	60
200	65
500	95

At what number of texts would the bill from each phone plan be the same? At what number of texts is Phone Company A the better choice? At what number of texts is Phone Company B the better choice?

Student solutions will vary. Sample solution is provided.

The equation that represents the function for Phone Company A is $y = 75$.

To determine the equation that represents the function for Phone Company B, we need the rate of change. (We are told it is constant.)

$$\frac{60 - 50}{150 - 50} = \frac{10}{100}$$
$$= 0.1$$

The equation for Phone Company B is shown below.

Using the assignment of 50 to 50,

$$50 = 0.1(50) + b$$
$$50 = 5 + b$$
$$45 = b.$$

The equation that represents the function for Phone Company B is $y = 0.1x + 45$.

We can determine at what point the phone companies charge the same amount by solving the system:

$$\begin{cases} y = 75 \\ y = 0.1x + 45 \end{cases}$$

$$75 = 0.1x + 45$$
$$30 = 0.1x$$
$$300 = x$$

After 300 texts are sent, both companies would charge the same amount, $75. More than 300 texts means that the bill from Phone Company B will be higher than Phone Company A. Less than 300 texts means the bill from Phone Company A will be higher.

EUREKA
MATH™

This work is derived from Eureka Math ™ and licensed by Great Minds. ©2015 Great Minds. eureka-math.org
G8-M5-TE-B4-1.3.0-07.2015

3. The function that gives the volume of water, y, that flows from Faucet A in gallons during x minutes is a linear function with the graph shown. Faucet B's water flow can be described by the equation $y = \frac{5}{6}x$, where y is the volume of water in gallons that flows from the faucet during x minutes. Assume the flow of water from each faucet is constant. Which faucet has a faster rate of flow of water? Each faucet is being used to fill a tub with a volume of 50 gallons. How long will it take each faucet to fill its tub? How do you know?

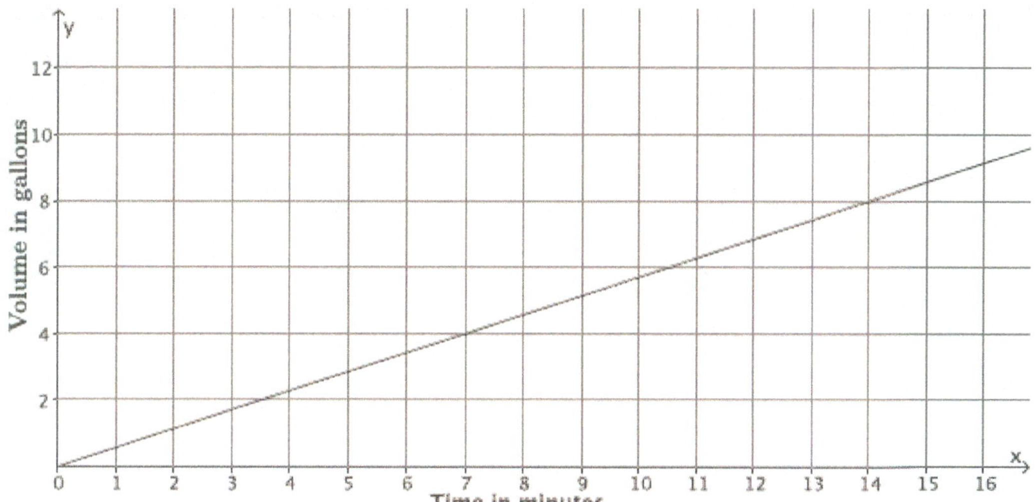

Suppose the tub being filled by Faucet A already had 15 gallons of water in it, and the tub being filled by Faucet B started empty. If now both faucets are turned on at the same time, which faucet will fill its tub fastest?

Student solutions will vary. Sample solution is provided.

The slope of the graph of the line is $\frac{4}{7}$ because $(7, 4)$ is a point on the line that represents 4 gallons of water that flows in 7 minutes. Therefore, the rate of water flow for Faucet A is $\frac{4}{7}$. To determine which faucet has a faster flow of water, we can compare their rates.

$$\frac{4}{7} < \frac{5}{6}$$

Therefore, Faucet B has a faster rate of water flow.

Faucet A	Faucet B	The tub filled by Faucet A that already has 15 gallons in it
$y = \frac{4}{7}x$ $50 = \frac{4}{7}x$ $50\left(\frac{7}{4}\right) = x$ $\frac{350}{4} = x$ $87.5 = x$ *It will take 87.5 minutes to fill a tub of 50 gallons.*	$y = \frac{5}{6}x$ $50 = \frac{5}{6}x$ $50\left(\frac{6}{5}\right) = x$ $60 = x$ *It will take 60 minutes to fill a tub of 50 gallons.*	$50 = \frac{4}{7}x + 15$ $35 = \frac{4}{7}x$ $35\left(\frac{7}{4}\right) = x$ $61.25 = x$ *Faucet B will fill the tub first because it will take Faucet A 61.25 minutes to fill the tub, even though it already has 15 gallons in it.*

This work is derived from Eureka Math ™ and licensed by Great Minds. ©2015 Great Minds. eureka-math.org
G8-M5-TE-B4-1.3.0-07.2015

4. Two people, Adam and Bianca, are competing to see who can save the most money in one month. Use the table and the graph below to determine who will save the most money at the end of the month. State how much money each person had at the start of the competition. (Assume each is following a linear function in his or her saving habit.)

Adam's Savings:

Bianca's Savings:

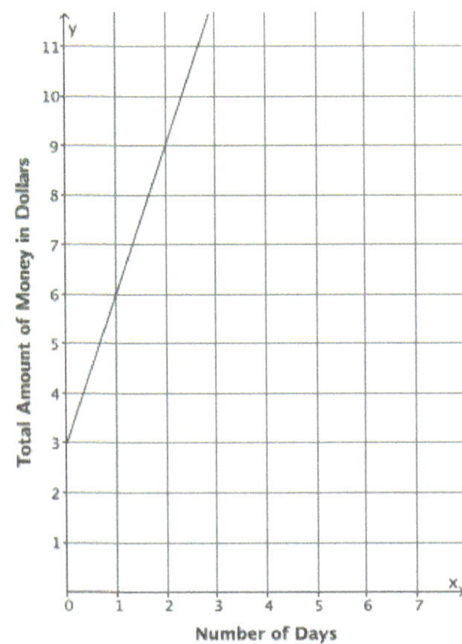

Input (Number of Days)	Output (Total amount of money in dollars)
5	17
8	26
12	38
20	62

The slope of the line that represents Adam's savings is 3; therefore, the rate at which Adam is saving money is $3 per day. According to the table of values for Bianca, she is also saving money at a rate of $3 per day:

$$\frac{26 - 17}{8 - 5} = \frac{9}{3} = 3$$

$$\frac{38 - 26}{12 - 8} = \frac{12}{4} = 3$$

$$\frac{62 - 26}{20 - 8} = \frac{36}{12} = 3$$

Therefore, at the end of the month, Adam and Bianca will both have saved the same amount of money.

According to the graph for Adam, the equation $y = 3x + 3$ represents the function of money saved each day. On day zero, he had $3.

The equation that represents the function of money saved each day for Bianca is $y = 3x + 2$ because, using the assignment of 17 to 5

$$17 = 3(5) + b$$
$$17 = 15 + b$$
$$2 = b.$$

The amount of money Bianca had on day zero was $2.

Lesson 7:   Comparing Linear Functions and Graphs

EUREKA MATH

This work is derived from Eureka Math ™ and licensed by Great Minds. ©2015 Great Minds. eureka-math.org
G8-M5-TE-B4-1.3.0-07.2015

Discussion (5 minutes)

To encourage students to compare different means of presenting linear functions, have students detail the different ways linear functions were described throughout these exercises. Use the following questions to guide the discussion.

MP.1

- Was one style of presentation easier to work with over the others? Does everyone agree?
- Was it easier to read off certain pieces of information about a linear function (its initial value, its constant rate of change, for instance) from one presentation of that function over another?

Closing (5 minutes)

Summarize, or ask students to summarize, the main points from the lesson:

- We know that functions can be expressed as equations, graphs, tables, and using verbal descriptions. Regardless of the way that a function is expressed, we can compare it with other functions.

Exit Ticket (5 minutes)

Fluency Exercise (10 minutes): Multi-Step Equations II

Rapid White Board Exchange (RWBE): During this exercise, students solve nine multi-step equations. Each equation should be solved in about a minute. Consider having students work on personal white boards, showing their solutions after each problem is assigned. The nine equations and their answers are at the end of the lesson. Refer to the Rapid White Board Exchanges section in the Module Overview for directions to administer a RWBE.

This work is derived from Eureka Math ™ and licensed by Great Minds. ©2015 Great Minds. eureka-math.org
G8-M5-TE-B4-1.3.0-07.2015

Name ____   Date _______________________

Lesson 7: Comparing Linear Functions and Graphs

Exit Ticket

Brothers Paul and Pete walk 2 miles to school from home. Paul can walk to school in 24 minutes. Pete has slept in again and needs to run to school. Paul walks at a constant rate, and Pete runs at a constant rate. The graph of the function that represents Pete's run is shown below.

a.   Which brother is moving at a greater rate? Explain how you know.

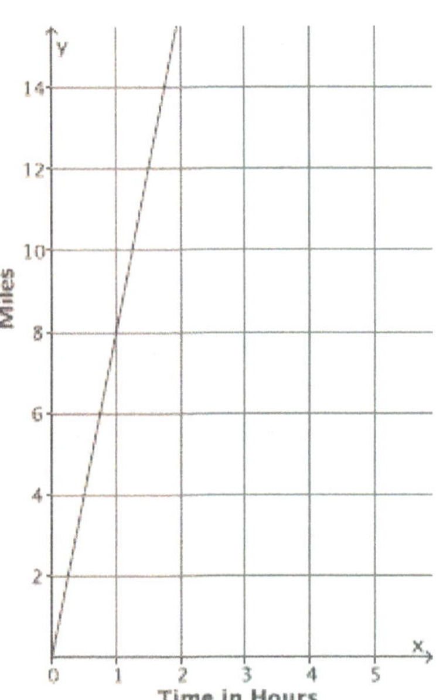

b.   If Pete leaves 5 minutes after Paul, will he catch up to Paul before they get to school?

Lesson 7:   Comparing Linear Functions and Graphs

EUREKA
MATH™

This work is derived from Eureka Math ™ and licensed by Great Minds. ©2015 Great Minds. eureka-math.org
G8-M5-TE-B4-1.3.0-07.2015

Exit Ticket Sample Solutions

Brothers Paul and Pete walk 2 miles to school from home. Paul can walk to school in 24 minutes. Pete has slept in again and needs to run to school. Paul walks at a constant rate, and Pete runs at a constant rate. The graph of the function that represents Pete's run is shown below.

a. **Which brother is moving at a greater rate? Explain how you know.**

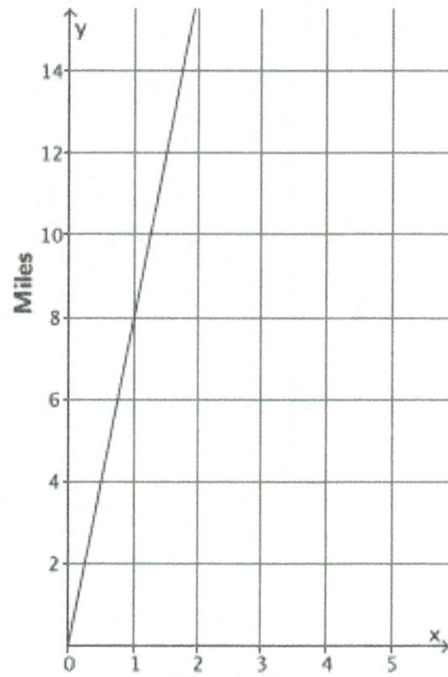

Paul takes 24 minutes to walk 2 miles; therefore, his rate is $\frac{1}{12}$ miles per minute.

Pete can run 8 miles in 60 minutes; therefore, his rate is $\frac{8}{60}$, or $\frac{2}{15}$ miles per minute.

Since $\frac{2}{15} > \frac{1}{12}$, Pete is moving at a greater rate.

b. **If Pete leaves 5 minutes after Paul, will he catch up to Paul before they get to school?**

Student solution methods will vary. Sample answer is shown.

Since Pete slept in, we need to account for that fact. So, Pete's time would be decreased. The equation that would represent the number of miles Pete runs, y, in x minutes, would be $y = \frac{2}{15}(x - 5)$.

The equation that would represent the number of miles Paul walks, y, in x minutes, would be $y = \frac{1}{12}x$.

To find out when they meet, solve the system of equations:

$$\begin{cases} y = \dfrac{2}{15}x - \dfrac{2}{3} \\ y = \dfrac{1}{12}x \end{cases}$$

$$\frac{2}{15}x - \frac{2}{3} = \frac{1}{12}x$$

$$\frac{2}{15}x - \frac{2}{3} - \frac{1}{12}x + \frac{2}{3} = \frac{1}{12}x - \frac{1}{12}x + \frac{2}{3}$$

$$\frac{1}{20}x = \frac{2}{3}$$

$$\left(\frac{20}{1}\right)\frac{1}{20}x = \frac{2}{3}\left(\frac{20}{1}\right)$$

$$x = \frac{40}{3}$$

$$y = \frac{1}{12}\left(\frac{40}{3}\right) = \frac{10}{9} \qquad \text{or} \qquad y = \frac{2}{15}\left(\frac{40}{3}\right) - \frac{2}{3}$$

Pete would catch up to Paul in $\frac{40}{3}$ minutes, which occurs $\frac{10}{9}$ miles from their home. Yes, he will catch Paul before they get to school because it is less than the total distance, two miles, to school.

EUREKA MATH™

This work is derived from Eureka Math ™ and licensed by Great Minds. ©2015 Great Minds. eureka-math.org
G8-M5-TE-B4-1.3.0-07.2015

Problem Set Sample Solutions

1. The graph below represents the distance in miles, y, Car A travels in x minutes. The table represents the distance in miles, y, Car B travels in x minutes. It is moving at a constant rate. Which car is traveling at a greater speed? How do you know?

Car A:

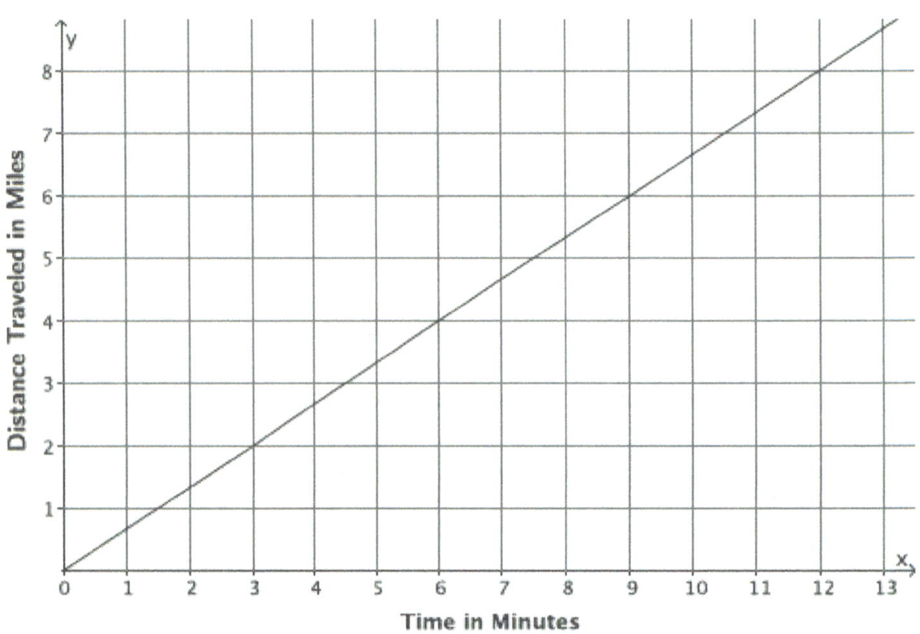

Car B:

Time in minutes (x)	Distance in miles (y)
15	12.5
30	25
45	37.5

Based on the graph, Car A is traveling at a rate of 2 miles every 3 minutes, $m = \frac{2}{3}$. From the table, the constant rate that Car B is traveling is

$$\frac{25 - 12.5}{30 - 15} = \frac{12.5}{15} = \frac{25}{30} = \frac{5}{6}.$$

Since $\frac{5}{6} > \frac{2}{3}$, Car B is traveling at a greater speed.

     Lesson 7:   Comparing Linear Functions and Graphs

This work is derived from Eureka Math ™ and licensed by Great Minds. ©2015 Great Minds. eureka-math.org
G8-M5-TE-B4-1.3.0-07.2015

2. The local park needs to replace an existing fence that is 6 feet high. Fence Company A charges $7,000 for building materials and $200 per foot for the length of the fence. Fence Company B charges are based solely on the length of the fence. That is, the total cost of the 6-foot high fence will depend on how long the fence is. The table below represents some inputs and their corresponding outputs that the cost function for Fence Company B assigns. It is a linear function.

Input (length of fence in feet)	Output (cost of bill in dollars)
100	26,000
120	31,200
180	46,800
250	65,000

a. Which company charges a higher rate per foot of fencing? How do you know?

Let x represent the length of the fence and y represent the total cost.

The equation that represents the function for Fence Company A is $y = 200x + 7,000$. So, the rate is 200 dollars per foot of fence.

The rate of change for Fence Company B is given by:

$$\frac{26,000 - 31,200}{100 - 120} = \frac{-5,200}{-20}$$
$$= 260$$

Fence Company B charges $260 per foot of fence, which is a higher rate per foot of fence length than Fence Company A.

b. At what number of the length of the fence would the cost from each fence company be the same? What will the cost be when the companies charge the same amount? If the fence you need were 190 feet in length, which company would be a better choice?

Student solutions will vary. Sample solution is provided.

The equation for Fence Company B is

$$y = 260x.$$

We can find out at what point the fence companies charge the same amount by solving the system

$$\begin{cases} y = 200x + 7000 \\ y = 260x \end{cases} \qquad \begin{aligned} 200x + 7,000 &= 260x \\ 7,000 &= 60x \\ 116.6666\ldots &= x \\ 116.7 &\approx x \end{aligned}$$

At 116.7 feet of fencing, both companies would charge the same amount (about $30,340$). Less than 116.7 feet of fencing means that the cost from Fence Company A will be more than Fence Company B. More than 116.7 feet of fencing means that the cost from Fence Company B will be more than Fence Company A. So, for 190 feet of fencing, Fence Company A is the better choice.

Lesson 7:   Comparing Linear Functions and Graphs

99

This work is derived from Eureka Math ™ and licensed by Great Minds. ©2015 Great Minds. eureka-math.org
G8-M5-TE-B4-1.3.0-07.2015

3. The equation $y = 123x$ describes the function for the number of toys, y, produced at Toys Plus in x minutes of production time. Another company, #1 Toys, has a similar function, also linear, that assigns the values shown in the table below. Which company produces toys at a slower rate? Explain.

Time in minutes (x)	Toys Produced (y)
5	600
11	1,320
13	1,560

We are told that #1 Toys produces toys at a constant rate. That rate is:

$$\frac{1,320 - 600}{11 - 5} = \frac{720}{6}$$
$$= 120$$

The rate of production for #1 Toys is 120 toys per minute. The rate of production for Toys Plus is 123 toys per minute. Since 120 is less than 123, #1 Toys produces toys at a slower rate.

4. A train is traveling from City A to City B, a distance of 320 miles. The graph below shows the number of miles, y, the train travels as a function of the number of hours, x, that have passed on its journey. The train travels at a constant speed for the first four hours of its journey and then slows down to a constant speed of 48 miles per hour for the remainder of its journey.

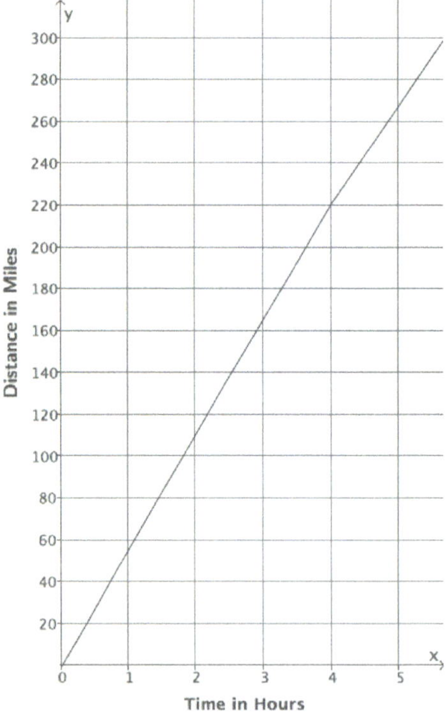

Lesson 7:   Comparing Linear Functions and Graphs

EUREKA MATH

This work is derived from Eureka Math ™ and licensed by Great Minds. ©2015 Great Minds. eureka-math.org
G8-M5-TE-B4-1.3.0-07.2015

a.   **How long will it take the train to reach its destination?**

Student solutions will vary. Sample solution is provided.

We see from the graph that the train travels 220 miles during its first four hours of travel. It has 100 miles remaining to travel, which it shall do at a constant speed of 48 miles per hour. We see that it will take about 2 hours more to finish the trip:

$$100 = 48x$$
$$2.08333 \ldots = x$$
$$2.1 \approx x.$$

This means it will take about 6.1 hours $(4 + 2.1 = 6.1)$ for the train to reach its destination.

b.   **If the train had not slowed down after 4 hours, how long would it have taken to reach its destination?**

$$320 = 55x$$
$$5.8181818 \ldots = x$$
$$5.8 \approx x$$

The train would have reached its destination in about 5.8 hours had it not slowed down.

c.   **Suppose after 4 hours, the train increased its constant speed. How fast would the train have to travel to complete the destination in 1.5 hours?**

Let m represent the new constant speed of the train.

$$100 = m(1.5)$$
$$66.6666 \ldots = x$$
$$66.7 \approx x$$

The train would have to increase its speed to about 66.7 miles per hour to arrive at its destination 1.5 hours later.

This work is derived from Eureka Math ™ and licensed by Great Minds. ©2015 Great Minds. eureka-math.org
G8-M5-TE-B4-1.3.0-07.2015

5.

a. A hose is used to fill up a $1,200$ gallon water truck. Water flows from the hose at a constant rate. After 10 minutes, there are 65 gallons of water in the truck. After 15 minutes, there are 82 gallons of water in the truck. How long will it take to fill up the water truck? Was the tank initially empty?

Student solutions will vary. Sample solution is provided.

Let x represent the time in minutes it takes to pump y gallons of water. Then, the rate can be found as follows:

Time in minutes (x)	Amount of water pumped in gallons (y)
10	65
15	82

$$\frac{65 - 82}{10 - 15} = \frac{-17}{-5}$$
$$= \frac{17}{5}$$

Since the water is pumping at a constant rate, we can assume the equation is linear. Therefore, the equation for the volume of water pumped from the hose is found by

$$65 = \frac{17}{5}(10) + b$$
$$65 = 34 + b$$
$$31 = b$$

The equation is $y = \frac{17}{5}x + 31$, and we see that the tank initially had 31 gallons of water in it. The time to fill the tank is given by

$$1200 = \frac{17}{5}x + 31$$
$$1169 = \frac{17}{5}x$$
$$343.8235 \ldots = x$$
$$343.8 \approx x$$

It would take about 344 minutes or about 5.7 hours to fill up the truck.

b. The driver of the truck realizes that something is wrong with the hose he is using. After 30 minutes, he shuts off the hose and tries a different hose. The second hose flows at a constant rate of 18 gallons per minute. How long now does it take to fill up the truck?

Since the first hose has been pumping for 30 minutes, there are 133 gallons of water already in the truck. That means the new hose only has to fill up $1,067$ gallons. Since the second hose fills up the truck at a constant rate of 18 gallons per minute, the equation for the second hose is $y = 18x$.

$$1067 = 18x$$
$$59.27 = x$$
$$59.3 \approx x$$

It will take the second hose about 59.3 minutes (or a little less than an hour) to finish the job.

Lesson 7:   Comparing Linear Functions and Graphs

EUREKA MATH

This work is derived from Eureka Math ™ and licensed by Great Minds. ©2015 Great Minds. eureka-math.org
G8-M5-TE-B4-1.3.0-07.2015

Multi-Step Equations II

1. $2(x + 5) = 3(x + 6)$

   $x = -8$

2. $3(x + 5) = 4(x + 6)$

   $x = -9$

3. $4(x + 5) = 5(x + 6)$

   $x = -10$

4. $-(4x + 1) = 3(2x - 1)$

   $x = \dfrac{1}{5}$

5. $3(4x + 1) = -(2x - 1)$

   $x = -\dfrac{1}{7}$

6. $-3(4x + 1) = 2x - 1$

   $x = -\dfrac{1}{7}$

7. $15x - 12 = 9x - 6$

   $x = 1$

8. $\dfrac{1}{3}(15x - 12) = 9x - 6$

   $x = \dfrac{1}{2}$

9. $\dfrac{2}{3}(15x - 12) = 9x - 6$

   $x = 2$

This work is derived from Eureka Math ™ and licensed by Great Minds. ©2015 Great Minds. eureka-math.org
G8-M5-TE-B4-1.3.0-07.2015

Lesson 8: Graphs of Simple Nonlinear Functions

Student Outcomes

- Students examine the average rate of change for nonlinear function over various intervals and verify that these values are not constant.

Lesson Notes

In Exercises 4–10, students are given the option to sketch the graphs of given equations to verify their claims about them being linear or nonlinear. For this reason, students may need graph paper to complete these exercises. Students need graph paper to complete the Problem Set.

Classwork

Exploratory Challenge/Exercises 1–3 (19 minutes)

Students work independently or in pairs to complete Exercises 1–3.

Exploratory Challenge/Exercises 1–3

1. Consider the function that assigns to each number x the value x^2.

   a. Do you think the function is linear or nonlinear? Explain.

   I think the function is nonlinear. The equation describing the function is not of the form $y = mx + b$.

   b. Develop a list of inputs and outputs for this function. Organize your work using the table below. Then, answer the questions that follow.

Input (x)	Output (x^2)
−5	25
−4	16
−3	9
−2	4
−1	1
0	0
1	1
2	4
3	9
4	16
5	25

Scaffolding:

Students may benefit from exploring these exercises in small groups.

Lesson 8:   Graphs of Simple Nonlinear Functions

EUREKA MATH™

This work is derived from Eureka Math ™ and licensed by Great Minds. ©2015 Great Minds. eureka-math.org
G8-M5-TE-B4-1.3.0-07.2015

c.   Plot the inputs and outputs as ordered pairs defining points on the coordinate plane.

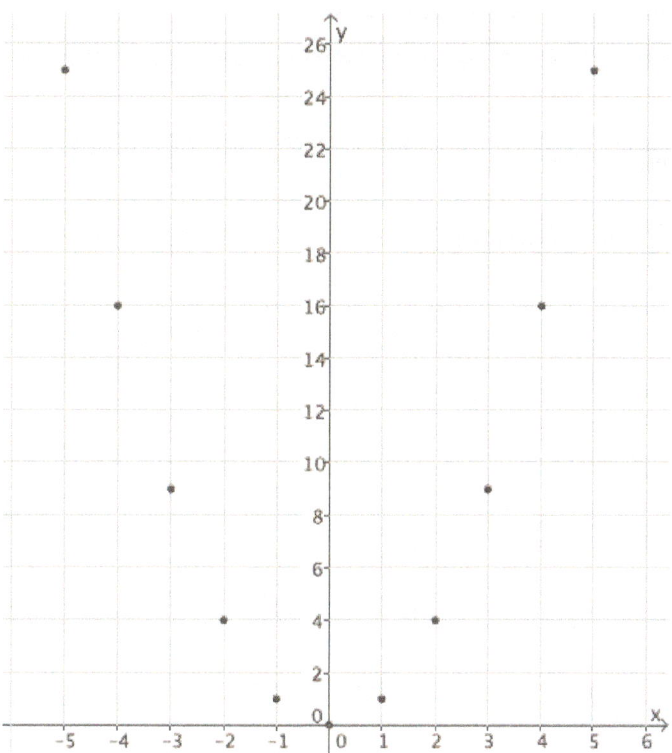

d.   What shape does the graph of the points appear to take?

It appears to take the shape of a curve.

e.   Find the rate of change using rows 1 and 2 from the table above.

$$\frac{25 - 16}{-5 - (-4)} = \frac{9}{-1} = -9$$

f.   Find the rate of change using rows 2 and 3 from the table above.

$$\frac{16 - 9}{-4 - (-3)} = \frac{7}{-1} = -7$$

g.   Find the rate of change using any two other rows from the table above.

Student work will vary.

$$\frac{16 - 25}{4 - 5} = \frac{-9}{-1} = 9$$

h.   Return to your initial claim about the function. Is it linear or nonlinear? Justify your answer with as many pieces of evidence as possible.

This is definitely a nonlinear function because the rate of change is not a constant for different intervals of inputs. Also, we would expect the graph of a linear function to be a set of points in a line, and this graph is not a line. As was stated before, the expression x^2 is nonlinear.

Lesson 8:   Graphs of Simple Nonlinear Functions

This work is derived from Eureka Math ™ and licensed by Great Minds. ©2015 Great Minds. eureka-math.org
G8-M5-TE-B4-1.3.0-07.2015

2. Consider the function that assigns to each number x the value x^3.

   a. Do you think the function is linear or nonlinear? Explain.

   I think the function is nonlinear. The equation describing the function is not of the form $y = mx + b$.

   b. Develop a list of inputs and outputs for this function. Organize your work using the table below. Then, answer the questions that follow.

Input (x)	Output (x^3)
-2.5	-15.625
-2	-8
-1.5	-3.375
-1	-1
-0.5	-0.125
0	0
0.5	0.125
1	1
1.5	3.375
2	8
2.5	15.625

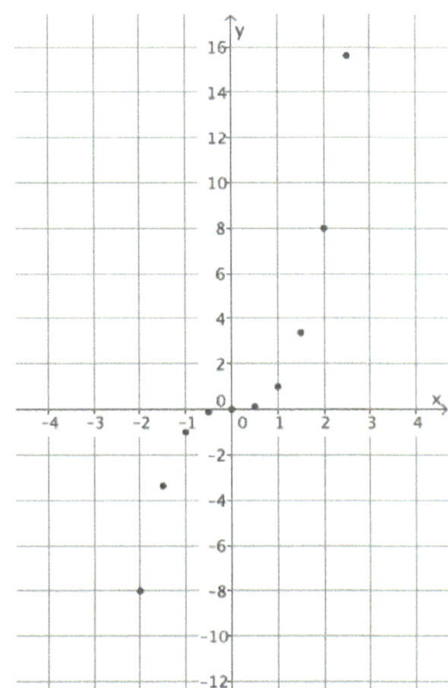

   c. Plot the inputs and outputs as ordered pairs defining points on the coordinate plane.

   d. What shape does the graph of the points appear to take?

   It appears to take the shape of a curve.

   e. Find the rate of change using rows 2 and 3 from the table above.

   $$\frac{-8 - (-3.375)}{-2 - (-1.5)} = \frac{-4.625}{-0.5} = 9.25$$

   f. Find the rate of change using rows 3 and 4 from the table above.

   $$\frac{-3.375 - (-1)}{-1.5 - (-1)} = \frac{-2.375}{-0.5} = 4.75$$

   g. Find the rate of change using rows 8 and 9 from the table above.

   $$\frac{1 - 3.375}{1 - 1.5} = \frac{-2.375}{-0.5} = 4.75$$

   h. Return to your initial claim about the function. Is it linear or nonlinear? Justify your answer with as many pieces of evidence as possible.

   This is definitely a nonlinear function because the rate of change is not a constant for any interval of inputs. Also, we would expect the graph of a linear function to be a line, and this graph is not a line. As was stated before, the expression x^3 is nonlinear.

Lesson 8:   Graphs of Simple Nonlinear Functions

This work is derived from Eureka Math ™ and licensed by Great Minds. ©2015 Great Minds. eureka-math.org
G8-M5-TE-B4-1.3.0-07.2015

EUREKA MATH™

3. Consider the function that assigns to each positive number x the value $\dfrac{1}{x}$.

   a. Do you think the function is linear or nonlinear? Explain.

   I think the function is nonlinear. The equation describing the function is not of the form $y = mx + b$.

   b. Develop a list of inputs and outputs for this function. Organize your work using the table below. Then, answer the questions that follow.

Input (x)	Output $\left(\dfrac{1}{x}\right)$
0.1	10
0.2	5
0.4	2.5
0.5	2
0.8	1.25
1	1
1.6	0.625
2	0.5
2.5	0.4
4	0.25
5	0.2

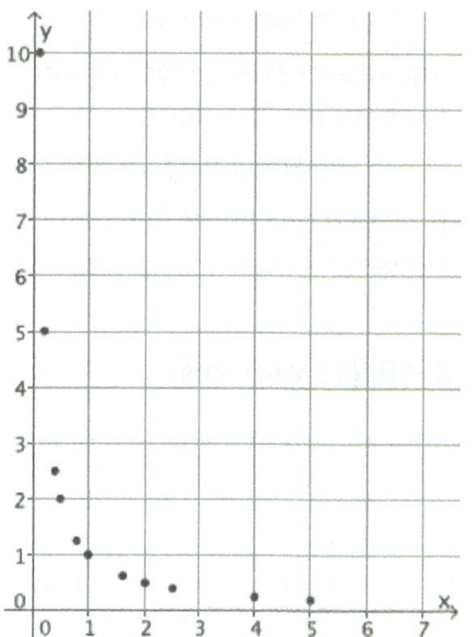

   c. Plot the inputs and outputs as ordered pairs defining points on the coordinate plane.

   d. What shape does the graph of the points appear to take?

   It appears to take the shape of a curve.

   e. Find the rate of change using rows 1 and 2 from the table above.

   $$\dfrac{10 - 5}{0.1 - 0.2} = \dfrac{5}{-0.1} = -50$$

   f. Find the rate of change using rows 2 and 3 from the table above.

   $$\dfrac{5 - 2.5}{0.2 - 0.4} = \dfrac{2.5}{-0.2} = -12.5$$

   g. Find the rate of change using any two other rows from the table above.

   Student work will vary.

   $$\dfrac{1 - 0.625}{1 - 1.6} = \dfrac{0.375}{-0.6} = -0.625$$

   h. Return to your initial claim about the function. Is it linear or nonlinear? Justify your answer with as many pieces of evidence as possible.

   This is definitely a nonlinear function because the rate of change is not a constant for any interval of inputs. Also, we would expect the graph of a linear function to be a line, and this graph is not a line. As was stated before, the expression $\dfrac{1}{x}$ is nonlinear.

This work is derived from Eureka Math ™ and licensed by Great Minds. ©2015 Great Minds. eureka-math.org
G8-M5-TE-B4-1.3.0-07.2015

Discussion (4 minutes)

- What did you notice about the rates of change in the preceding three problems?
    - *The rates of change were not constant in each of the three problems.*
- If the rate of change for pairs of inputs and corresponding outputs were the same for each and every pair, then what can we say about the function?
    - *We know the function is linear.*
- If the rate of change for pairs of inputs and corresponding outputs is not the same for each pair, what can we say about the function?
    - *We know the function is nonlinear.*
- Recall that any linear function can be described by an equation of the form $y = mx + b$. Any equation that cannot be written in this form is not linear, and its corresponding function is nonlinear.

Exercises 4–10 (12 minutes)

Students work independently or in pairs to complete Exercises 4–10.

Exercises 4–10

In each of Exercises 4–10, an equation describing a rule for a function is given, and a question is asked about it. If necessary, use a table to organize pairs of inputs and outputs, and then plot each on a coordinate plane to help answer the question.

4. What shape do you expect the graph of the function described by $y = x$ to take? Is it a linear or nonlinear function?

    I expect the shape of the graph to be a line. This function is a linear function described by the linear equation $y = x$. The graph of this function is a line.

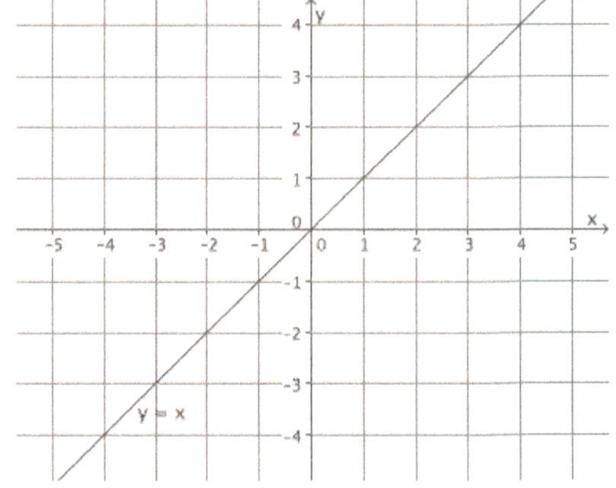

5. What shape do you expect the graph of the function described by $y = 2x^2 - x$ to take? Is it a linear or nonlinear function?

    I expect the shape of the graph to be something other than a line. This function is nonlinear because its graph is not a line. Also the equation describing the function is not of the form $y = mx + b$. It is not linear.

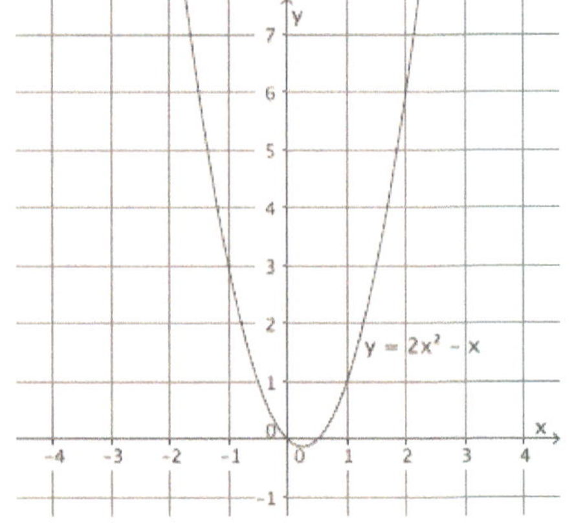

Lesson 8:   Graphs of Simple Nonlinear Functions

This work is derived from Eureka Math ™ and licensed by Great Minds. ©2015 Great Minds. eureka-math.org
G8-M5-TE-B4-1.3.0-07.2015

6. What shape do you expect the graph of the function described by $3x + 7y = 8$ to take? Is it a linear or nonlinear function?

I expect the shape of the graph to be a line. This function is a linear function described by the linear equation $3x + 7y = 8$. The graph of this function is a line. (We have $y = -\frac{3}{7}x + \frac{8}{7}$.)

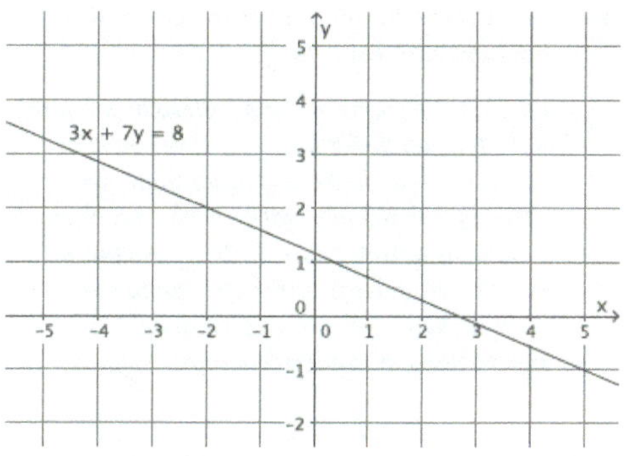

7. What shape do you expect the graph of the function described by $y = 4x^3$ to take? Is it a linear or nonlinear function?

I expect the shape of the graph to be something other than a line. This function is nonlinear because its graph is not a line. Also the equation describing the function is not of the form $y = mx + b$. It is not linear.

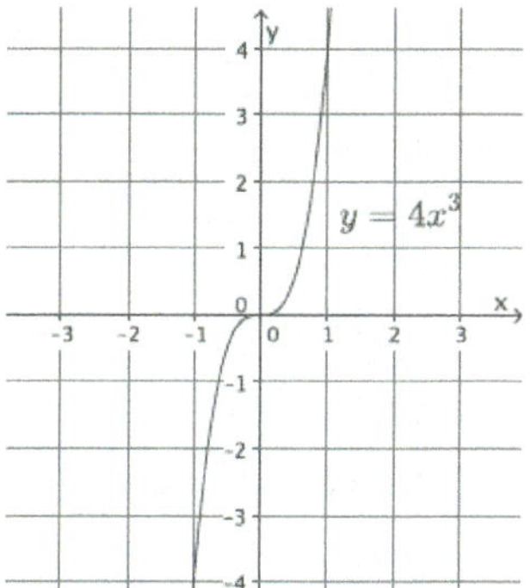

8. What shape do you expect the graph of the function described by $\frac{3}{x} = y$ to take? Is it a linear or nonlinear function? (Assume that an input of $x = 0$ is disallowed.)

I expect the shape of the graph to be something other than a line. This function is nonlinear because its graph is not a line. Also the equation describing the function is not of the form $y = mx + b$. It is not linear.

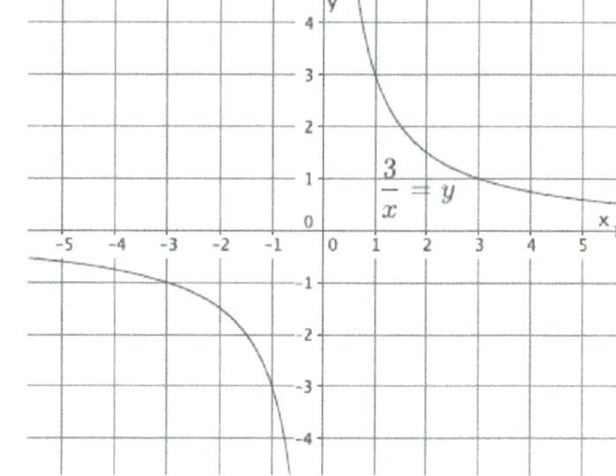

9. What shape do you expect the graph of the function described by $\frac{4}{x^2} = y$ to take? Is it a linear or nonlinear function? (Assume that an input of $x = 0$ is disallowed.)

I expect the shape of the graph to be something other than a line. This function is nonlinear because its graph is not a line. Also the equation describing the function is not of the form $y = mx + b$. It is not linear.

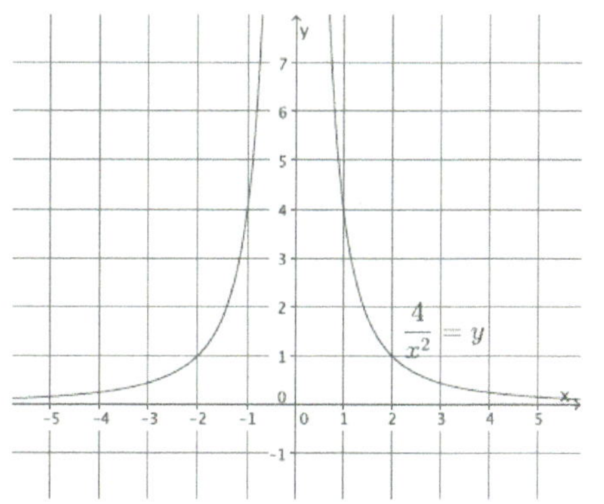

This work is derived from Eureka Math ™ and licensed by Great Minds. ©2015 Great Minds. eureka-math.org
G8-M5-TE-B4-1.3.0-07.2015

10. What shape do you expect the graph of the equation $x^2 + y^2 = 36$ to take? Is it a linear or nonlinear function? Is it a function? Explain.

I expect the shape of the graph to be something other than a line. It is nonlinear because its graph is not a line. It is not a function because there is more than one output for any given value of x in the interval $(-6, 6)$. For example, at $x = 0$ the y-value is both 6 and -6. This does not fit the definition of function because functions assign to each input exactly one output. Since there is at least one instance where an input has two outputs, it is not a function.

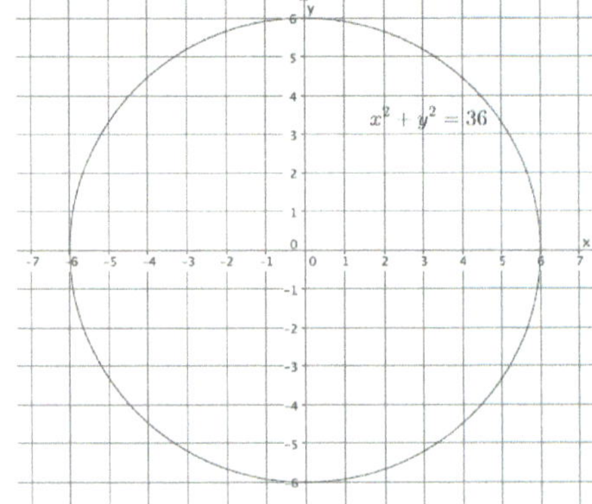

Closing (5 minutes)

Summarize, or ask students to summarize, the main points from the lesson.

- Students understand that, unlike linear functions, nonlinear functions do not have a constant rate of change.
- Students expect the graph of nonlinear functions to be some sort of curve.

Lesson Summary

One way to determine if a function is linear or nonlinear is to inspect average rates of change using a table of values. If these average rates of change are not constant, then the function is not linear.

Another way is to examine the graph of the function. If all the points on the graph do not lie on a common line, then the function is not linear.

If a function is described by an equation different from one equivalent to $y = mx + b$ for some fixed values m and b, then the function is not linear.

Exit Ticket (5 minutes)

Lesson 8:   Graphs of Simple Nonlinear Functions

This work is derived from Eureka Math ™ and licensed by Great Minds. ©2015 Great Minds. eureka-math.org
G8-M5-TE-B4-1.3.0-07.2015

Name ____   Date ____________________

Lesson 8: Graphs of Simple Nonlinear Functions

Exit Ticket

1. The graph below is the graph of a function. Do you think the function is linear or nonlinear? Briefly justify your answer.

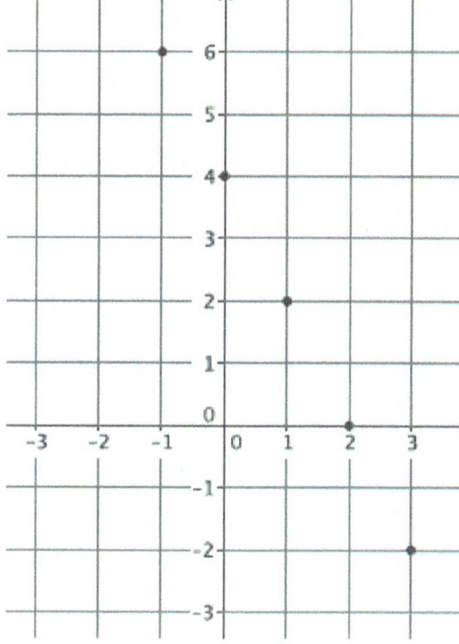

2. Consider the function that assigns to each number x the value $\frac{1}{2}x^2$. Do you expect the graph of this function to be a straight line? Briefly justify your answer.

EUREKA MATH

Lesson 8:   Graphs of Simple Nonlinear Functions

111

This work is derived from Eureka Math ™ and licensed by Great Minds. ©2015 Great Minds. eureka-math.org
G8-M5-TE-B4-1.3.0-07.2015

Exit Ticket Sample Solutions

1. The graph below is the graph of a function. Do you think the function is linear or nonlinear? Briefly justify your answer.

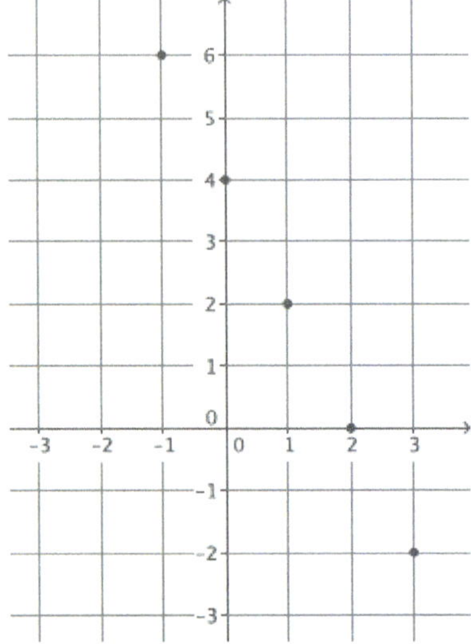

Student work may vary. The plot of this graph appears to be a straight line, and so the function is linear.

2. Consider the function that assigns to each number x the value $\frac{1}{2}x^2$. Do you expect the graph of this function to be a straight line? Briefly justify your answer.

The equation is nonlinear (not of the form $y = mx + b$), so the function is nonlinear. Its graph will not be a straight line.

Problem Set Sample Solutions

1. Consider the function that assigns to each number x the value $x^2 - 4$.

   a. Do you think the function is linear or nonlinear? Explain.

   The equation describing the function is not of the form $y = mx + b$. It is not linear.

   b. Do you expect the graph of this function to be a straight line?

   No

Lesson 8:   Graphs of Simple Nonlinear Functions

EUREKA MATH

This work is derived from Eureka Math ™ and licensed by Great Minds. ©2015 Great Minds. eureka-math.org
G8-M5-TE-B4-1.3.0-07.2015

c.   Develop a list of inputs and matching outputs for this function. Use them to begin a graph of the function.

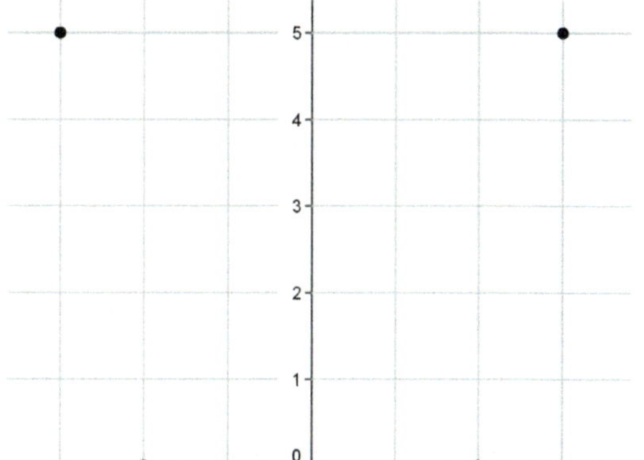

Input (x)	Output ($x^2 - 4$)
-3	5
-2	0
-1	-3
0	-4
1	-3
2	0
3	5

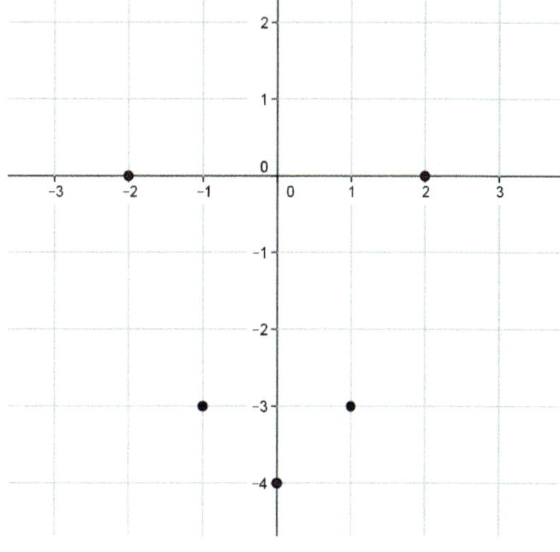

d.   Was your prediction to (b) correct?

Yes, the graph appears to be taking the shape of some type of curve.

2.   Consider the function that assigns to each number x greater than -3 the value $\frac{1}{x+3}$.

a.   Is the function linear or nonlinear? Explain.

The equation describing the function is not of the form $y = mx + b$. It is not linear.

b.   Do you expect the graph of this function to be a straight line?

No

c.   Develop a list of inputs and matching outputs for this function. Use them to begin a graph of the function.

Input (x)	Output ($\frac{1}{x+3}$)
-2	1
-1	0.5
0	$0.3333\ldots$
1	0.25
2	0.2
3	$0.16666\ldots$

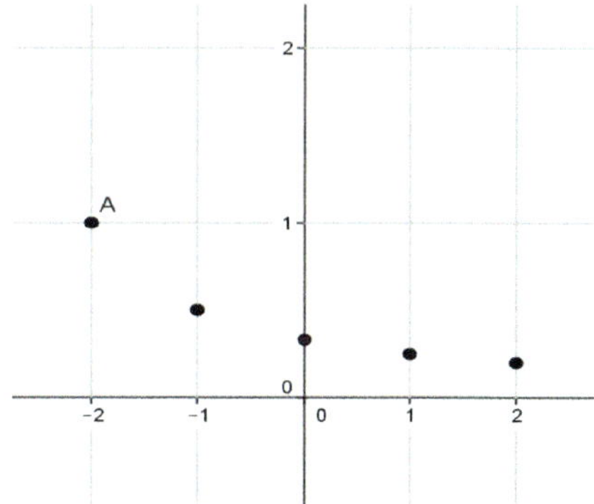

d.   Was your prediction to (b) correct?

Yes, the graph appears to be taking the shape of some type of curve.

This work is derived from Eureka Math ™ and licensed by Great Minds. ©2015 Great Minds. eureka-math.org
G8-M5-TE-B4-1.3.0-07.2015

3.

a. Is the function represented by this graph linear or nonlinear? Briefly justify your answer.

The graph is clearly not a straight line, so the function is not linear.

b. What is the average rate of change for this function from an input of $x = -2$ to an input of $x = -1$?

$$\frac{-2 - 1}{-2 - (-1)} = \frac{-3}{-1} = 3$$

c. What is the average rate of change for this function from an input of $x = -1$ to an input of $x = 0$?

$$\frac{1 - 2}{-1 - 0} = \frac{-1}{-1} = 1$$

As expected, the average rate of change of this function is not constant.

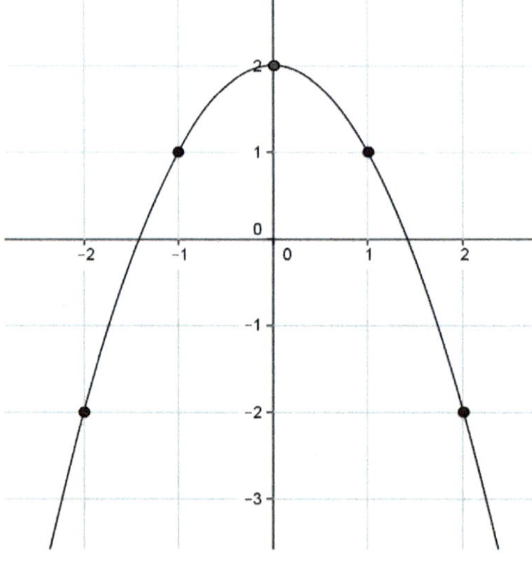

Lesson 8:   Graphs of Simple Nonlinear Functions

This work is derived from Eureka Math ™ and licensed by Great Minds. ©2015 Great Minds. eureka-math.org
G8-M5-TE-B4-1.3.0-07.2015

EUREKA
MATH™

Mathematics Curriculum

8 GRADE

Topic B

Volume

8.G.C.9

Focus Standard:	8.G.C.9	Know the formulas for the volumes of cones, cylinders, and spheres and use them to solve real-world and mathematical problems.
Instructional Days:	3	
Lesson 9:	Examples of Functions from Geometry (E)[1]	
Lesson 10:	Volumes of Familiar Solids—Cones and Cylinders (S)	
Lesson 11:	Volume of a Sphere (P)	

In Lesson 9, students work with functions from geometry. For example, students write the rules that represent the perimeters of various regular shapes and areas of common shapes. Along those same lines, students write functions that represent the area of more complex shapes (e.g., the border of a picture frame). In Lesson 10, students learn the volume formulas for cylinders and cones. Building upon their knowledge of area of circles and the concept of congruence, students see a cylinder as a stack of circular congruent disks and consider the total area of the disks in three dimensions as the volume of a cylinder. A physical demonstration shows students that it takes exactly three cones of the same dimensions as a cylinder to equal the volume of the cylinder. The demonstration leads students to the formula for the volume of cones in general. Students apply the formulas to answer questions such as, "If a cone is filled with water to half its height, what is the ratio of the volume of water to the container itself?" Students then use what they know about the volume of the cylinder to derive the formula for the volume of a sphere. In Lesson 11, students learn that the volume of a sphere is equal to two-thirds the volume of a cylinder that fits tightly around the sphere and touches only at points. Finally, students apply what they have learned about volume to solve real-world problems where they will need to make decisions about which formulas to apply to a given situation.

---

[1]Lesson Structure Key: **P**-Problem Set Lesson, **M**-Modeling Cycle Lesson, **E**-Exploration Lesson, **S**-Socratic Lesson

This work is derived from Eureka Math ™ and licensed by Great Minds. ©2015 Great Minds. eureka-math.org
G8-M5-TE-B4-1.3.0-07.2015

Lesson 9: Examples of Functions from Geometry

Student Outcomes

- Students write rules to express functions related to geometry.
- Students review what they know about volume with respect to rectangular prisms and further develop their conceptual understanding of volume by comparing the liquid contained within a solid to the volume of a standard rectangular prism (i.e., a prism with base area equal to one).

Classwork

Exploratory Challenge 1/Exercises 1–4 (10 minutes)

Students work independently or in pairs to complete Exercises 1–4. Once students are finished, debrief the activity. Ask students to think about real-life situations that might require using the function they developed in Exercise 4. Some sample responses may include area of wood needed to make a 1-inch frame for a picture, area required to make a sidewalk border (likely larger than 1-inch) around a park or playground, or the area of a planter around a tree.

Exploratory Challenge 1/Exercises 1–4

As you complete Exercises 1–4, record the information in the table below.

	Side length in inches (s)	Area in square inches (A)	Expression that describes area of border
Exercise 1	6	36	$64 - 36$
	8	64	
Exercise 2	9	81	$121 - 81$
	11	121	
Exercise 3	13	169	$225 - 169$
	15	225	
Exercise 4	s	s^2	$(s + 2)^2 - s^2$
	$s + 2$	$(s + 2)^2$	

This work is derived from Eureka Math ™ and licensed by Great Minds. ©2015 Great Minds. eureka-math.org
G8-M5-TE-B4-1.3.0-07.2015

EUREKA MATH™

1. **Use the figure below to answer parts (a)–(f).**

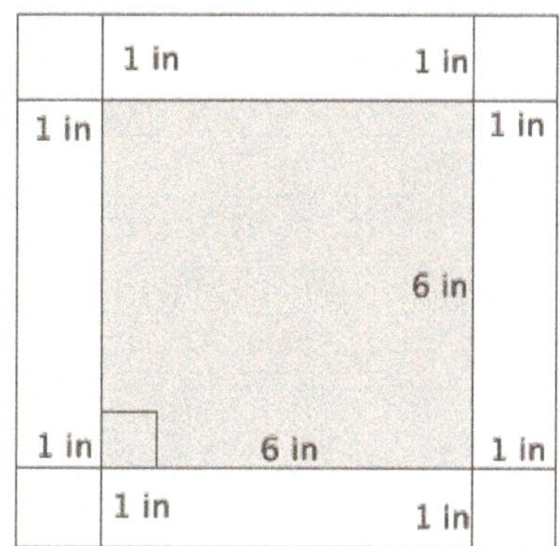

a. **What is the length of one side of the smaller, inner square?**

    The length of one side of the smaller square is 6 in.

b. **What is the area of the smaller, inner square?**

    $6^2 = 36$

    The area of the smaller square is 36 in^2.

c. **What is the length of one side of the larger, outer square?**

    The length of one side of the larger square is 8 in.

d. **What is the area of the larger, outer square?**

    $8^2 = 64$

    The area of the larger square is 64 in^2.

e. **Use your answers in parts (b) and (d) to determine the area of the 1-inch white border of the figure.**

    $64 - 36 = 28$

    The area of the 1-inch white border is 28 in^2.

f. **Explain your strategy for finding the area of the white border.**

    First, I had to determine the length of one side of the larger, outer square. Since the inner square is 6 in. and the border is 1 in. on all sides, then the length of one side of the larger square is $(6 + 2)$ in $= 8$ in. Then, the area of the larger square is 64 in^2. Next, I found the area of the smaller, inner square. Since one side length is 6 in., the area is 36 in^2. To find the area of the white border, I needed to subtract the area of the inner square from the area of the outer square.

This work is derived from Eureka Math ™ and licensed by Great Minds. ©2015 Great Minds. eureka-math.org
G8-M5-TE-B4-1.3.0-07.2015

2. **Use the figure below to answer parts (a)–(f).**

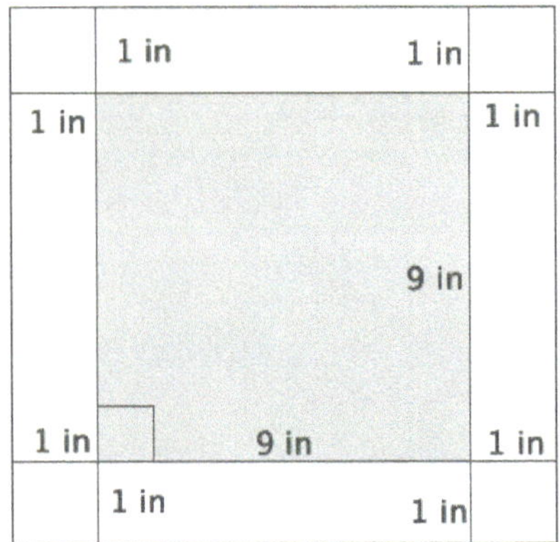

a. **What is the length of one side of the smaller, inner square?**

The length of one side of the smaller square is 9 *in.*

b. **What is the area of the smaller, inner square?**

$9^2 = 81$

The area of the smaller square is 81 *in^2.*

c. **What is the length of one side of the larger, outer square?**

The length of one side of the larger square is 11 *in.*

d. **What is the area of the larger, outer square?**

$11^2 = 121$

The area of the larger square is 121 *in^2.*

e. **Use your answers in parts (b) and (d) to determine the area of the 1-inch white border of the figure.**

$121 - 81 = 40$

The area of the 1-inch white border is 40 *in^2.*

f. **Explain your strategy for finding the area of the white border.**

First, I had to determine the length of one side of the larger, outer square. Since the inner square is 9 *in. and the border is* 1 *in. on all sides, the length of one side of the larger square is* $(9 + 2)$ *in* $= 11$ *in. Therefore, the area of the larger square is* 121 *in^2. Then, I found the area of the smaller, inner square. Since one side length is* 9 *in., the area is* 81 *in^2. To find the area of the white border, I needed to subtract the area of the inner square from the area of the outer square.*

Lesson 9:   Examples of Functions from Geometry

This work is derived from Eureka Math ™ and licensed by Great Minds. ©2015 Great Minds. eureka-math.org
G8-M5-TE-B4-1.3.0-07.2015

EUREKA
MATH™

3. Use the figure below to answer parts (a)–(f).

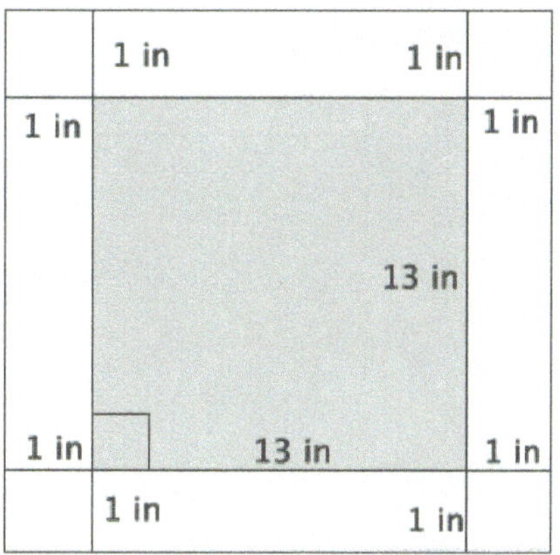

a. What is the length of one side of the smaller, inner square?

The length of one side of the smaller square is 13 in.

b. What is the area of the smaller, inner square?

$13^2 = 169$

The area of the smaller square is 169 in^2.

c. What is the length of one side of the larger, outer square?

The length of one side of the larger square is 15 in.

d. What is the area of the larger, outer square?

$15^2 = 225$

The area of the larger square is 225 in^2.

e. Use your answers in parts (b) and (d) to determine the area of the 1-inch white border of the figure.

$225 - 169 = 56$

The area of the 1-inch white border is 56 in^2.

f. Explain your strategy for finding the area of the white border.

First, I had to determine the length of one side of the larger, outer square. Since the inner square is 13 in. and the border is 1 in. on all sides, the length of one side of the larger square is $(13 + 2)$ in = 15 in. Therefore, the area of the larger square is 225 in^2. Then, I found the area of the smaller, inner square. Since one side length is 13 in., the area is 169 in^2. To find the area of the white border, I needed to subtract the area of the inner square from the area of the outer square.

This work is derived from Eureka Math ™ and licensed by Great Minds. ©2015 Great Minds. eureka-math.org
G8-M5-TE-B4-1.3.0-07.2015

4. Write a function that would allow you to calculate the area of a 1-inch white border for any sized square picture measured in inches.

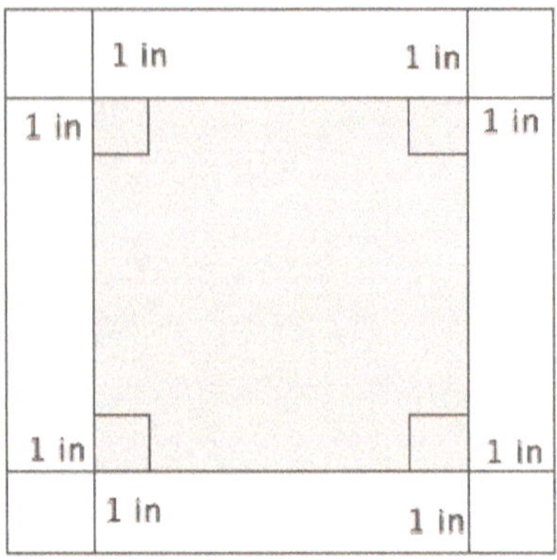

a. Write an expression that represents the side length of the smaller, inner square.

Symbols used will vary. Expect students to use s or x to represent one side of the smaller, inner square. Answers that follow will use s as the symbol to represent one side of the smaller, inner square.

b. Write an expression that represents the area of the smaller, inner square.

$$s^2$$

c. Write an expression that represents the side lengths of the larger, outer square.

$$s + 2$$

d. Write an expression that represents the area of the larger, outer square.

$$(s + 2)^2$$

e. Use your expressions in parts (b) and (d) to write a function for the area A of the 1-inch white border for any sized square picture measured in inches.

$$A = (s + 2)^2 - s^2$$

Discussion (6 minutes)

This discussion prepares students for the volume problems that they will work in the next two lessons. The goal is to remind students of the concept of volume using a rectangular prism and then have them describe the volume in terms of a function.

- Recall the concept of volume. How do you describe the volume of a three-dimensional figure? Give an example, if necessary.

  □ *Volume is the space that a three-dimensional figure can occupy. The volume of a glass is the amount of liquid it can hold.*

- In Grade 6 you learned the formula to determine the volume of a rectangular prism. The volume V of a rectangular prism is a function of the edge lengths, l, w, and h. That is, the function that allows us to determine the volume of a rectangular prism can be described by the following rule:

$$V = lwh.$$

Lesson 9:   Examples of Functions from Geometry

This work is derived from Eureka Math ™ and licensed by Great Minds. ©2015 Great Minds. eureka-math.org
G8-M5-TE-B4-1.3.0-07.2015

- Generally, we interpret volume in the following way:
- Fill the shell of the solid with water, and pour water into a three-dimensional figure, in this case a standard rectangular prism (i.e., a prism with base side lengths of one), as shown.

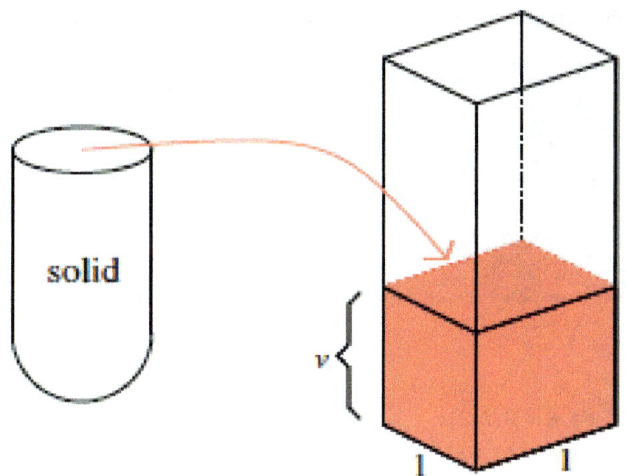

> **Scaffolding:**
> - Concrete and hands-on experiences with volume would be useful.
> - Students may know the formulas for volume but with different letters to represent the values (linked to their first language).

- Then, the volume of the shell of the solid is the height v of the water in the standard rectangular prism. Why is the volume, v, the height of the water?

  □ *The volume is equal to the height of the water because the area of the base is 1 square unit. Thus, whatever the height, v, is, multiplied by 1, will be equal to v.*

- If the height of water in the standard rectangular prism is 16.7 ft., what is the volume of the shell of the solid? Explain.

  □ *The volume of the shell of the solid would be 16.7 ft^3 because the height, 16.7 ft., multiplied by the area of the base, 1 ft^2, is 16.7 ft^3.*

- There are a few basic assumptions that we make when we discuss volume. Have students paraphrase each assumption after you state it to make sure they understand the concept.

(a) The volume of a solid is always a number greater than or equal to 0.

(b) The volume of a unit cube (i.e., a rectangular prism whose edges all have length 1) is by definition 1 cubic unit.

(c) If two solids are identical, then their volumes are equal.

(d) If two solids have (at most) their boundaries in common, then their total volume can be calculated by adding the individual volumes together. (These figures are sometimes referred to as composite solids.)

This work is derived from Eureka Math ™ and licensed by Great Minds. ©2015 Great Minds. eureka-math.org
G8-M5-TE-B4-1.3.0-07.2015

Exercises 5–6 (5 minutes)

Exercises 5–6

5. The volume of the prism shown below is 61.6 in^3. What is the height of the prism?

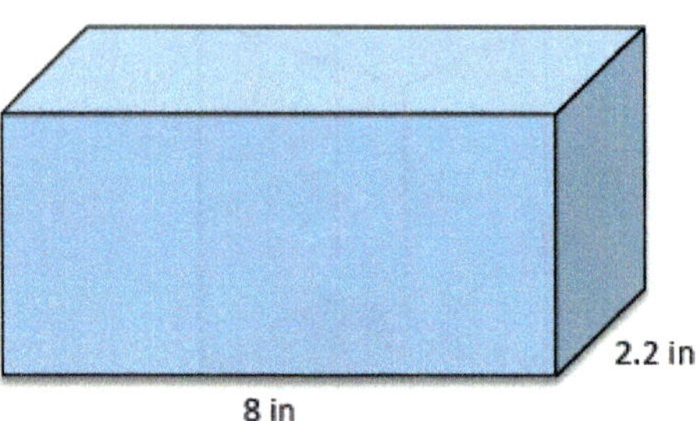

Let x represent the height of the prism.

$$61.6 = 8(2.2)x$$
$$61.6 = 17.6x$$
$$3.5 = x$$

The height of the prism is 3.5 in.

6. Find the value of the ratio that compares the volume of the larger prism to the smaller prism.

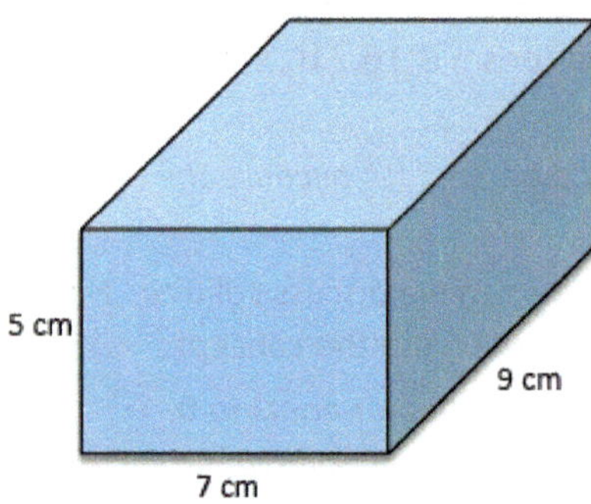

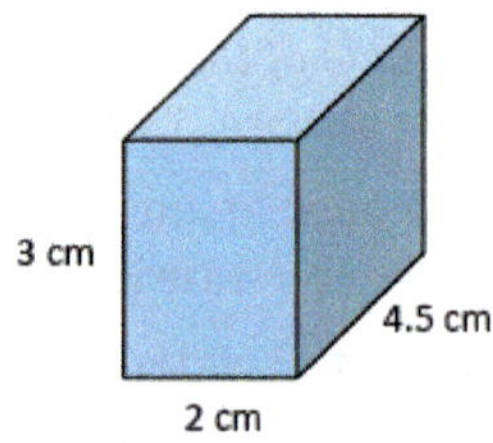

Volume of larger prism:

$$V = 7(9)(5)$$
$$= 315$$

The volume of the larger prism is 315 cm^3.

Volume of smaller prism:

$$V = 2(4.5)(3)$$
$$= 27$$

The volume of the smaller prism is 27 cm^3.

The ratio that compares the volume of the larger prism to the smaller prism is $315\!:\!27$. The value of the ratio is $\dfrac{315}{27} = \dfrac{35}{3}$.

Lesson 9:   Examples of Functions from Geometry

EUREKA MATH™

This work is derived from Eureka Math ™ and licensed by Great Minds. ©2015 Great Minds. eureka-math.org
G8-M5-TE-B4-1.3.0-07.2015

Exploratory Challenge 2/Exercises 7–10 (14 minutes)

Students work independently or in pairs to complete Exercises 7–10. Ensure that students know that when *base* is referenced, it means the bottom of the prism.

Exploratory Challenge 2/Exercises 7–10

As you complete Exercises 7–10, record the information in the table below. Note that *base* refers to the bottom of the prism.

	Area of base in square centimeters (B)	Height in centimeters (h)	Volume in cubic centimeters
Exercise 7	36	3	108
Exercise 8	36	8	288
Exercise 9	36	15	540
Exercise 10	36	x	$36x$

7. Use the figure to the right to answer parts (a)–(c).

    a. What is the area of the base?

    The area of the base is 36 *cm*^{2}.

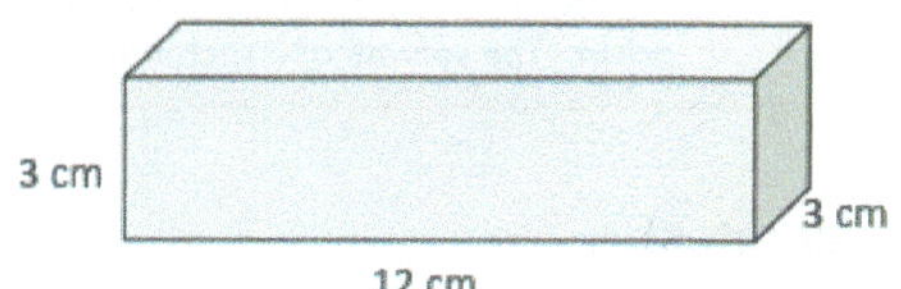

    b. What is the height of the figure?

    The height is 3 *cm.*

    c. What is the volume of the figure?

    The volume of the rectangular prism is 108 *cm*^{3}.

8. Use the figure to the right to answer parts (a)–(c).

    a. What is the area of the base?

    The area of the base is 36 *cm*^{2}.

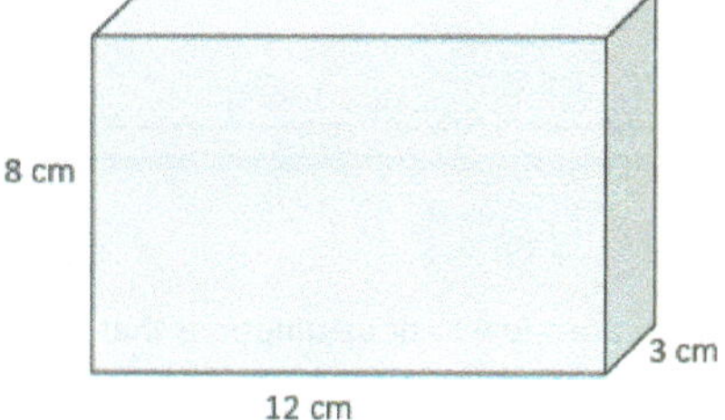

    b. What is the height of the figure?

    The height is 8 *cm.*

    c. What is the volume of the figure?

    The volume of the rectangular prism is 288 *cm*^{3}.

9. Use the figure to the right to answer parts (a)–(c).

    a. What is the area of the base?

    The area of the base is 36 *cm*^{2}.

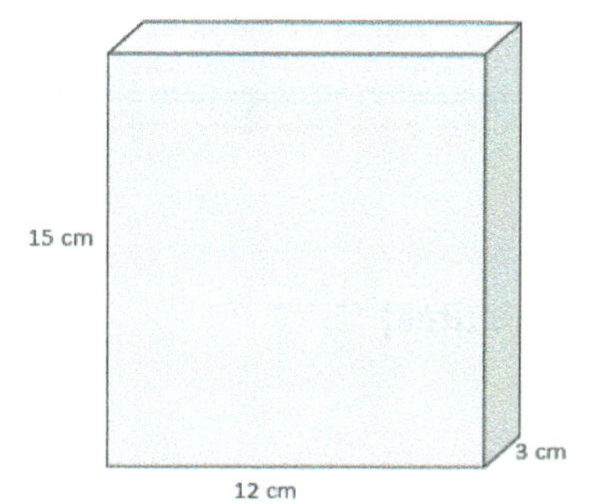

    b. What is the height of the figure?

    The height is 15 *cm.*

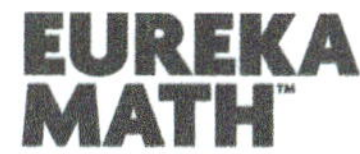

This work is derived from Eureka Math ™ and licensed by Great Minds. ©2015 Great Minds. eureka-math.org
G8-M5-TE-B4-1.3.0-07.2015

c.   What is the volume of the figure?

The volume of the rectangular prism is 540 cm^3.

10. Use the figure to the right to answer parts (a)–(c).

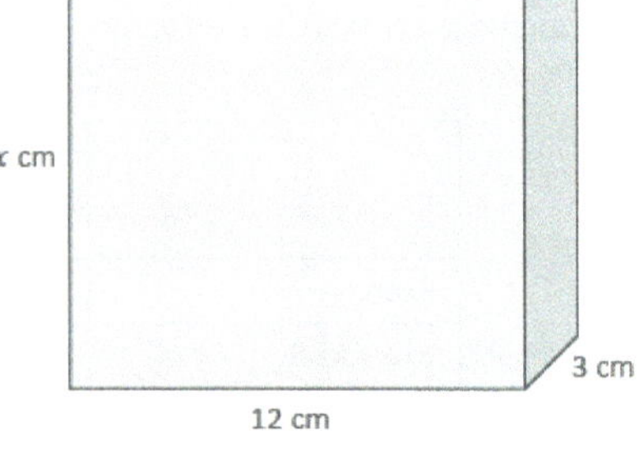

a.   What is the area of the base?

The area of the base is 36 cm^2.

b.   What is the height of the figure?

The height is x *cm.*

MP.8

c.   Write and describe a function that will allow you to determine the volume of any rectangular prism that has a base area of 36 cm^2.

The rule that describes the function is $V = 36x$, *where* V *is the volume and* x *is the height of the rectangular prism. The volume of a rectangular prism with a base area of* 36 cm^2 *is a function of its height.*

Closing (5 minutes)

Summarize, or ask students to summarize, the main points from the lesson:

- We know how to write functions to determine area or volume of a figure.
- We know that we can add volumes together as long as they only touch at a boundary.
- We know that identical solids will be equal in volume.
- We were reminded of the volume formula for a rectangular prism, and we used the formula to determine the volume of rectangular prisms.

Lesson Summary

There are a few basic assumptions that are made when working with volume:

(a)   The volume of a solid is always a number greater than or equal to 0.

(b)   The volume of a unit cube (i.e., a rectangular prism whose edges all have a length of 1) is by definition 1 cubic unit.

(c)   If two solids are identical, then their volumes are equal.

(d)   If two solids have (at most) their boundaries in common, then their total volume can be calculated by adding the individual volumes together. (These figures are sometimes referred to as composite solids.)

Exit Ticket (5 minutes)

Lesson 9:   Examples of Functions from Geometry

This work is derived from Eureka Math ™ and licensed by Great Minds. ©2015 Great Minds. eureka-math.org
G8-M5-TE-B4-1.3.0-07.2015

Name ____   Date _______________________

Lesson 9: Examples of Functions from Geometry

Exit Ticket

1. Write a function that would allow you to calculate the area in square inches, A, of a 2-inch white border for any sized square figure with sides of length s measured in inches.

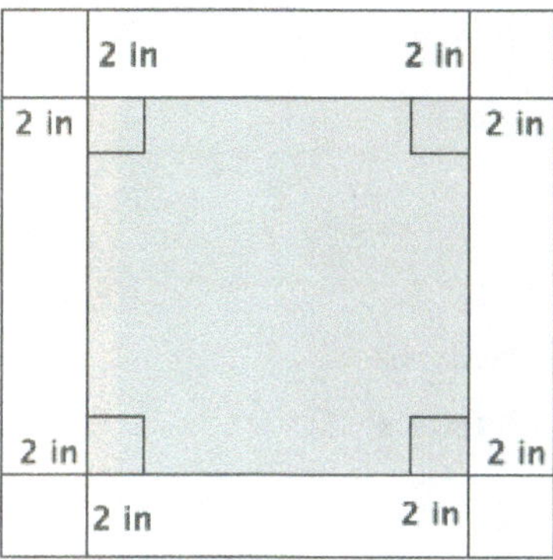

2. The volume of the rectangular prism is 295.68 in^3. What is its width?

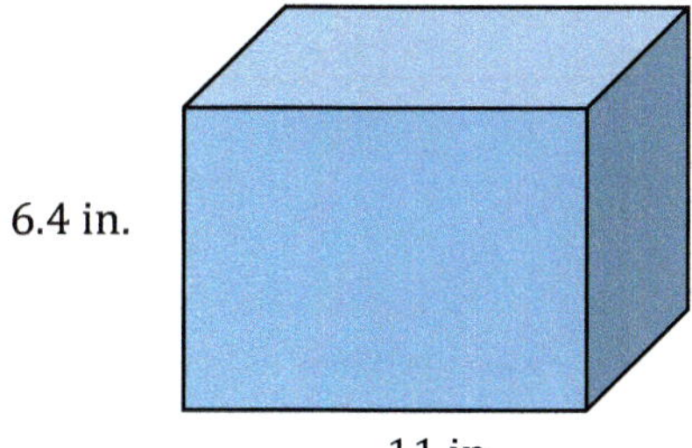

This work is derived from Eureka Math ™ and licensed by Great Minds. ©2015 Great Minds. eureka-math.org
G8-M5-TE-B4-1.3.0-07.2015

Exit Ticket Sample Solutions

1. Write a function that would allow you to calculate the area in square inches, A, of a 2-inch white border for any sized square figure with sides of length s measured in inches.

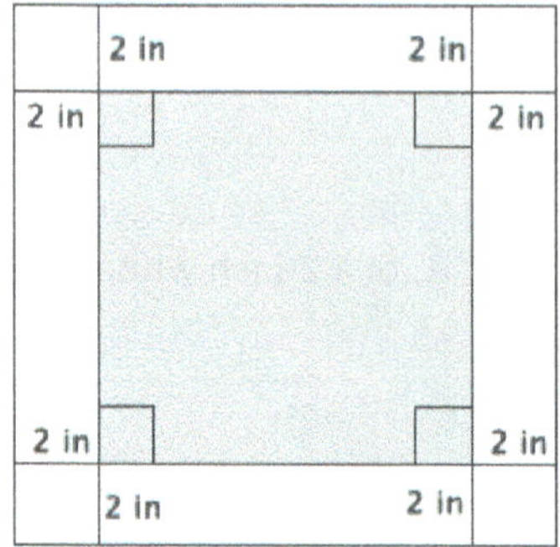

Let s represent the side length of the inner square in inches. Then, the area of the inner square is s^2 square inches. The side length of the larger square, in inches, is $s + 4$, and the area in square inches is $(s + 4)^2$. If A is the area of the 2-inch border, then the function that describes A in square inches is

$$A = (s + 4)^2 - s^2.$$

2. The volume of the rectangular prism is 295.68 in^3. What is its width?

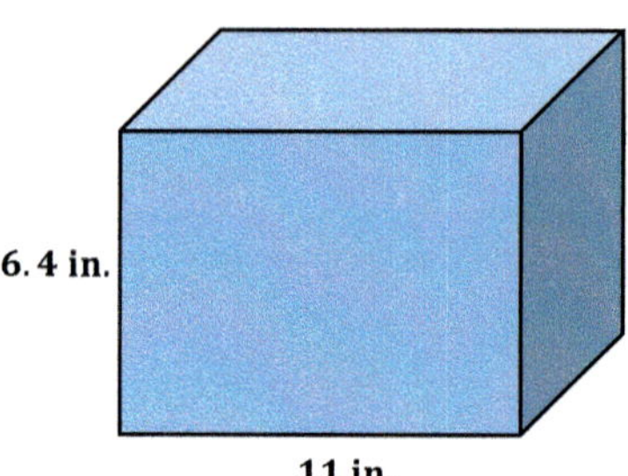

Let x represent the width of the prism.

$$295.68 = 11(6.4)x$$
$$295.68 = 70.4x$$
$$4.2 = x$$

The width of the prism is 4.2 in.

Problem Set Sample Solutions

1. Calculate the area of the 3-inch white border of the square figure below.

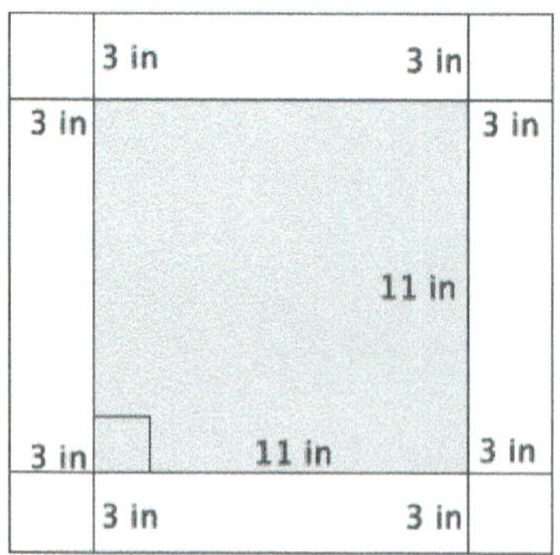

$$17^2 = 289$$
$$11^2 = 121$$

The area of the 3-inch white border is 168 in^2.

Lesson 9:   Examples of Functions from Geometry

EUREKA MATH

This work is derived from Eureka Math ™ and licensed by Great Minds. ©2015 Great Minds. eureka-math.org
G8-M5-TE-B4-1.3.0-07.2015

2. Write a function that would allow you to calculate the area, A, of a 3-inch white border for any sized square picture measured in inches.

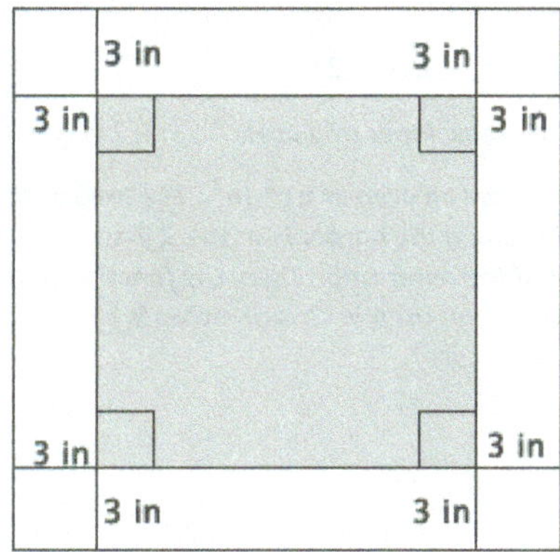

Let s represent the side length of the inner square in inches. Then, the area of the inner square is s^2 square inches. The side length of the outer square, in inches, is $s + 6$, which means that the area of the outer square, in square inches, is $(s + 6)^2$. The function that describes the area, A, of the 3-inch border is in square inches

$$A = (s + 6)^2 - s^2.$$

3. Dartboards typically have an outer ring of numbers that represent the number of points a player can score for getting a dart in that section. A simplified dartboard is shown below. The center of the circle is point A. Calculate the area of the outer ring. Write an exact answer that uses π (*do not* approximate your answer by using 3.14 for π).

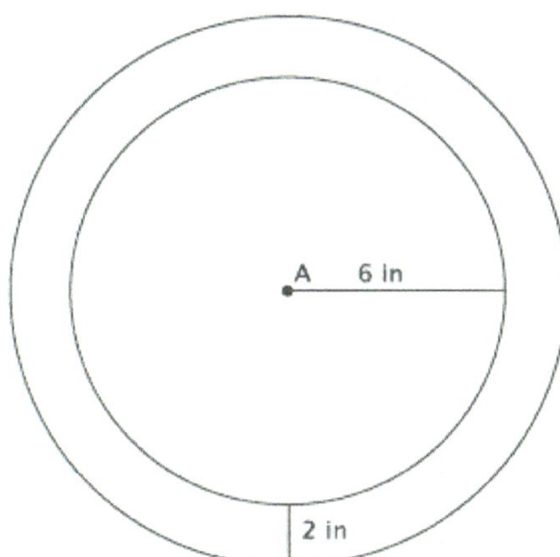

Inner ring area: $\pi r^2 = \pi(6^2) = 36\,\pi$

Outer ring: $\pi r^2 = \pi(6 + 2)^2 = \pi(8^2) = 64\,\pi$

Difference in areas: $64\,\pi - 36\,\pi = (64 - 34)\pi = 28\,\pi$

The inner ring has an area of 36π in^2. The area of the inner ring including the border is 64π in^2. The difference is the area of the border, 28π in^2.

EUREKA MATH™

This work is derived from Eureka Math ™ and licensed by Great Minds. ©2015 Great Minds. eureka-math.org
G8-M5-TE-B4-1.3.0-07.2015

4. Write a function that would allow you to calculate the area, A, of the outer ring for any sized dartboard with radius r. Write an exact answer that uses π (*do not* approximate your answer by using 3.14 for π).

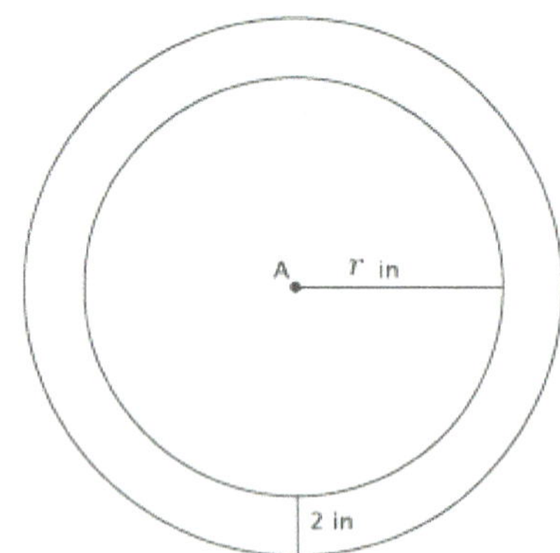

Inner ring area: πr^2

Outer ring: $\pi r^2 = \pi(r+2)^2$

Difference in areas: Inner ring area: $\pi(r+2)^2 - \pi r^2$

The inner ring has an area of πr^2 in². The area of the inner ring including the border is $\pi(r+2)^2$ in². Let A be the area of the outer ring. Then, the function that would describe that area in square inches is
$A = \pi(r+2)^2 - \pi r^2$.

5. The shell of the solid shown was filled with water and then poured into the standard rectangular prism, as shown. The height that the volume reaches is 14.2 in. What is the volume of the shell of the solid?

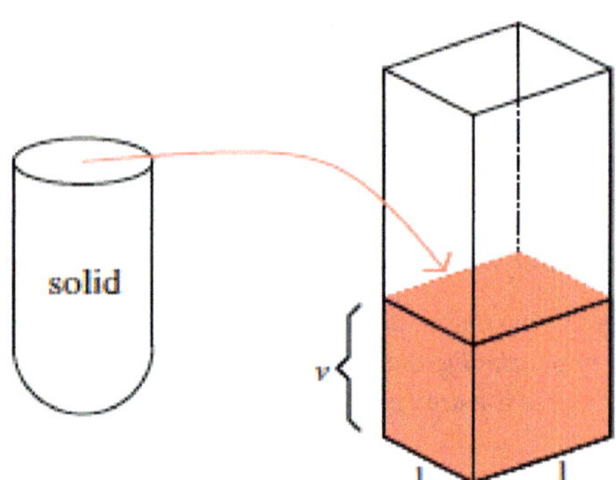

$$V = Bh$$
$$= 1(14.2)$$
$$= 14.2$$

The volume of the shell of the solid is 14.2 in³.

6. Determine the volume of the rectangular prism shown below.

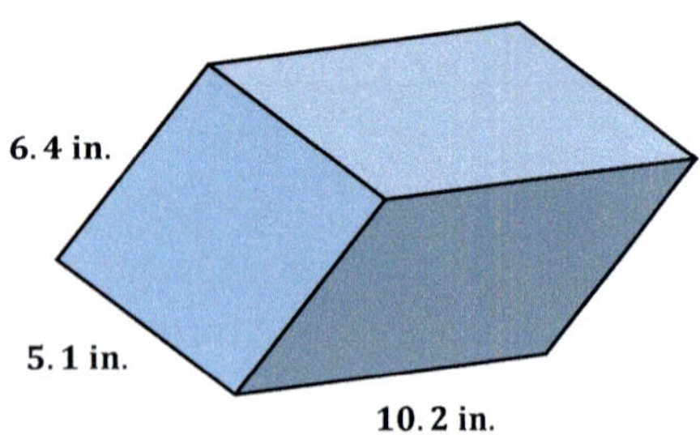

$6.4 \times 5.1 \times 10.2 = 332.928$

The volume of the prism is 332.928 in³.

EUREKA MATH™

This work is derived from Eureka Math ™ and licensed by Great Minds. ©2015 Great Minds. eureka-math.org
G8-M5-TE-B4-1.3.0-07.2015

7. The volume of the prism shown below is 972 cm³. What is its length?

Let x represent the length of the prism.

$$972 = 8.1(5)x$$
$$972 = 40.5x$$
$$24 = x$$

The length of the prism is 24 cm.

8. The volume of the prism shown below is 32.7375 ft³. What is its width?

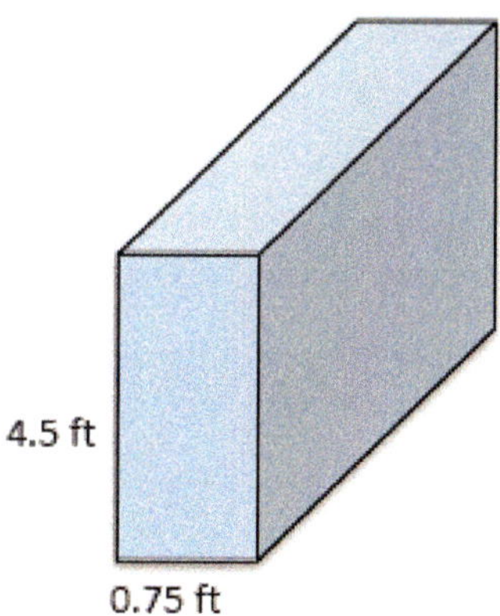

Let x represent the width.

$$32.7375 = (0.75)(4.5)x$$
$$32.7375 = 3.375x$$
$$9.7 = x$$

The width of the prism is 9.7 ft.

9. Determine the volume of the three-dimensional figure below. Explain how you got your answer.

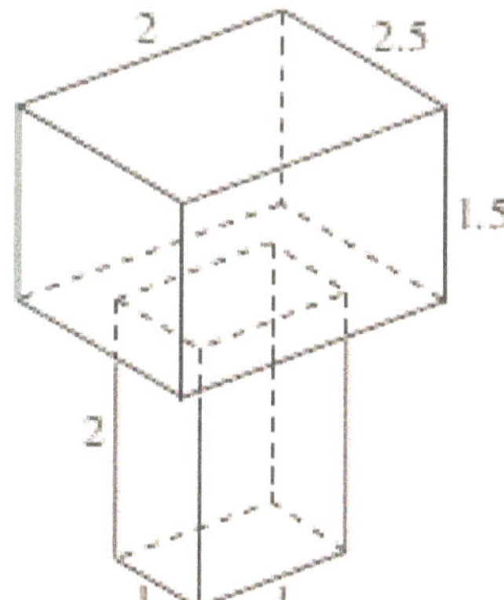

$$2 \times 2.5 \times 1.5 = 7.5$$
$$2 \times 1 \times 1 = 2$$

The volume of the top rectangular prism is 7.5 units³.
The volume of the bottom rectangular prism is 2 units³.
The figure is made of two rectangular prisms, and since the
rectangular prisms only touch at their boundaries, we can
add their volumes together to obtain the volume of the
figure. The total volume of the three-dimensional figure is
9.5 units³.

EUREKA MATH™

This work is derived from Eureka Math ™ and licensed by Great Minds. ©2015 Great Minds. eureka-math.org
G8-M5-TE-B4-1.3.0-07.2015

Lesson 10: Volumes of Familiar Solids—Cones and Cylinders

Student Outcomes

- Students know the volume formulas for cones and cylinders.
- Students apply the formulas for volume to real-world and mathematical problems.

Lesson Notes

For the demonstrations in this lesson, the following items are needed: a stack of same-sized note cards, a stack of same-sized round disks, a cylinder and cone of the same dimensions, and something with which to fill the cone (e.g., rice, sand, or water). Demonstrate to students that the volume of a rectangular prism is like finding the sum of the areas of congruent rectangles, stacked one on top of the next. A similar demonstration is useful for the volume of a cylinder. To demonstrate that the volume of a cone is one-third that of the volume of a cylinder with the same dimension, fill a cone with rice, sand, or water, and show students that it takes exactly three cones to equal the volume of the cylinder.

Classwork

Opening Exercise (3 minutes)

Students complete the Opening Exercise independently. Revisit the Opening Exercise once the discussion below is finished.

Opening Exercise

a.

    i.   **Write an equation to determine the volume of the rectangular prism shown below.**

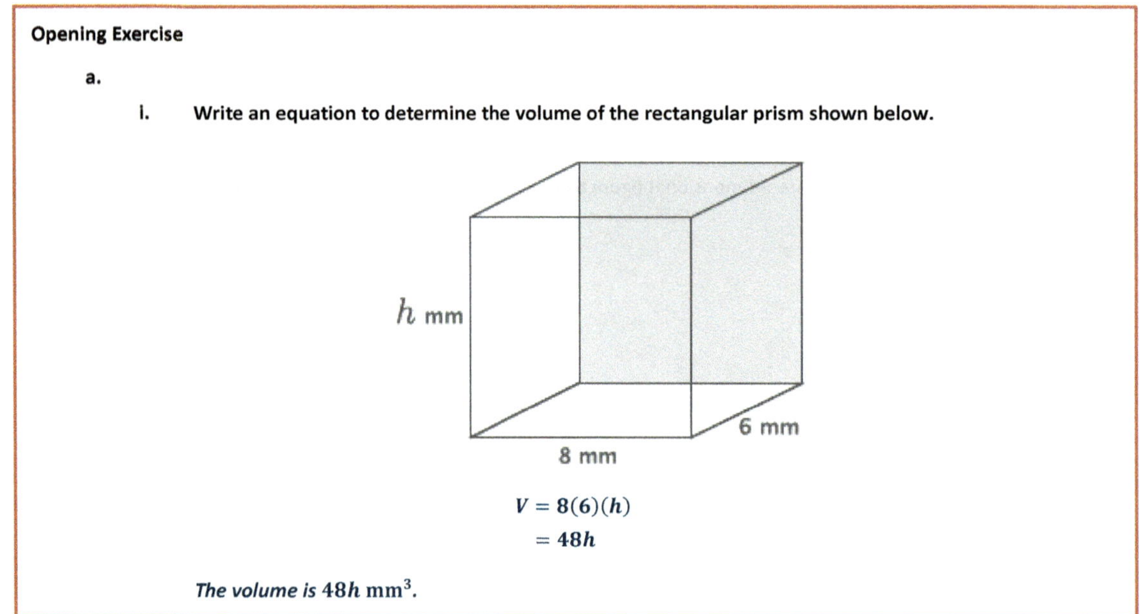

$$V = 8(6)(h)$$
$$= 48h$$

The volume is $48h$ mm^3.

EUREKA MATH™

This work is derived from Eureka Math ™ and licensed by Great Minds. ©2015 Great Minds. eureka-math.org
G8-M5-TE-B4-1.3.0-07.2015

ii.   Write an equation to determine the volume of the rectangular prism shown below.

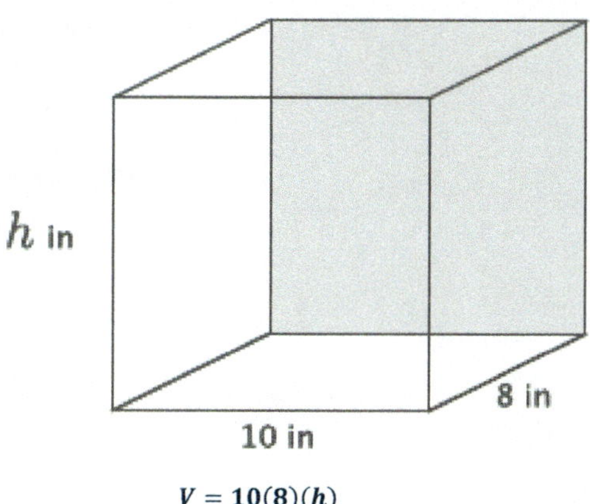

$$V = 10(8)(h)$$
$$= 80h$$

The volume is $80h$ in^3.

iii.   Write an equation to determine the volume of the rectangular prism shown below.

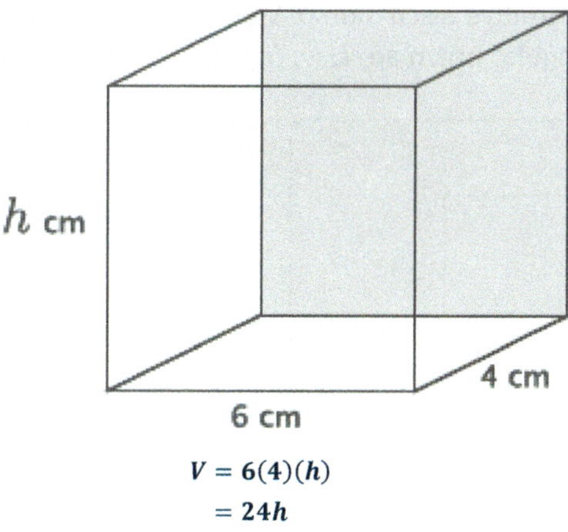

$$V = 6(4)(h)$$
$$= 24h$$

The volume is $24h$ cm^3.

iv.   Write an equation for volume, V, in terms of the area of the base, B.

$$V = Bh$$

Lesson 10:   Volumes of Familiar Solids—Cones and Cylinders

131

This work is derived from Eureka Math ™ and licensed by Great Minds. ©2015 Great Minds. eureka-math.org
G8-M5-TE-B4-1.3.0-07.2015

b.   Using what you learned in part (a), write an equation to determine the volume of the cylinder shown below.

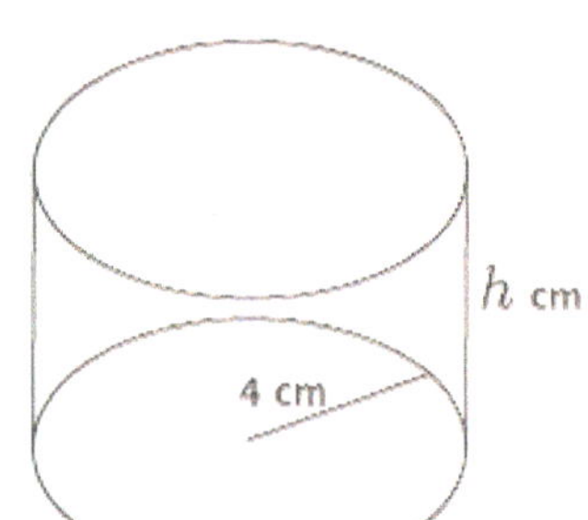

$$V = Bh$$
$$= 4^2 \pi h$$
$$= 16 \pi h$$

The volume is $16\pi h$ cm^3.

MP.1
&
MP.7

Students may not know the formula to determine the volume of a cylinder, so some may not be able to respond to this exercise until after the discussion below. This is an exercise for students to make sense of problems and persevere in solving them.

Discussion (10 minutes)

- We will continue with an intuitive discussion of volume. The volume formula from the last lesson says that if the dimensions of a rectangular prism are l, w, h, then the volume of the rectangular prism is $V = l \cdot w \cdot h$.

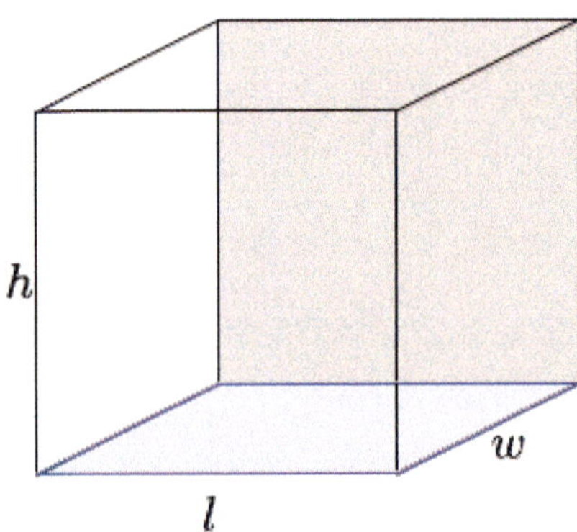

> *Scaffolding:*
>
> Demonstrate the volume of a rectangular prism using a stack of note cards. The volume of the rectangular prism increases as the height of the stack increases. Note that the rectangles (note cards) are congruent.

- Referring to the picture, we call the blue rectangle at the bottom of the rectangular prism the *base*, and the length of any one of the edges perpendicular to the base the *height* of the rectangular prism. Then, the formula says

$$V = (\text{area of base}) \cdot \text{height}.$$

    Lesson 10:   Volumes of Familiar Solids—Cones and Cylinders

This work is derived from Eureka Math ™ and licensed by Great Minds. ©2015 Great Minds. eureka-math.org
G8-M5-TE-B4-1.3.0-07.2015

- Examine the volume of a cylinder with base B and height h. Is the solid (i.e., the totality of all the line segments) of length h lying above the plane so that each segment is perpendicular to the plane, and is its lower endpoint lying on the base B (as shown)?

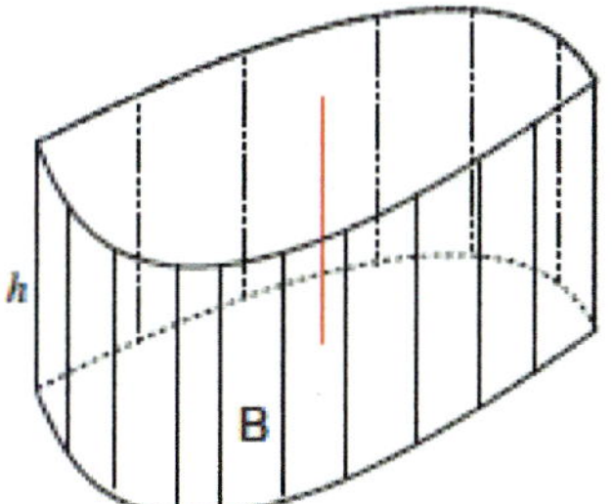

> *Scaffolding:*
>
> Clearly stating the meanings of symbols may present challenges for English language learners, and as such, students may benefit from a menu of phrases to support their statements. They will require detailed instruction and support in learning the non-negotiable vocabulary terms and phrases.

- Do you know a name for the shape of the base?

    □ *No, it is some curvy shape.*

- Let's examine another cylinder.

- Do we know the name of the shape of the base?

    □ *It appears to be a circle.*

- What do you notice about the line segments intersecting the base?

    □ *The line segments appear to be perpendicular to the base.*

- What angle does the line segment appear to make with the base?

    □ *The angle appears to be a right angle.*

- When the base of a diagram is the shape of a circle and the line segments on the base are perpendicular to the base, then the shape of the diagram is called a *right circular cylinder*.

We want to use the general formula for volume of a prism to apply to this shape of a right circular cylinder.

> *Scaffolding:*
>
> Demonstrate the volume of a cylinder using a stack of round disks. The volume of the cylinder increases as the height of the stack increases, just like the rectangular prism. Note that the disks are congruent.

- What is the general formula for finding the volume of a prism?

    □ $V = $ (area of base) $\cdot$ height

- What is the area for the base of the right circular cylinder?

    □ *The area of a circle is $A = \pi r^2$.*

MP.8

- What information do we need to find the area of a circle?

    □ *We need to know the radius of the circle.*

- What would be the volume of a right circular cylinder?

    □ $V = (\pi r^2)h$

- What information is needed to find the volume of a right circular cylinder?

    □ *We would need to know the radius of the base and the height of the cylinder.*

Lesson 10:   Volumes of Familiar Solids—Cones and Cylinders

133

This work is derived from Eureka Math ™ and licensed by Great Minds. ©2015 Great Minds. eureka-math.org
G8-M5-TE-B4-1.3.0-07.2015

Exercises 1–3 (8 minutes)

Students work independently or in pairs to complete Exercises 1–3.

Exercises 1–3

1. Use the diagram to the right to answer the questions.

   a. What is the area of the base?

   The area of the base is $(4.5)(8.2)$ *in*^{2} *or* 36.9 *in*^{2}.

   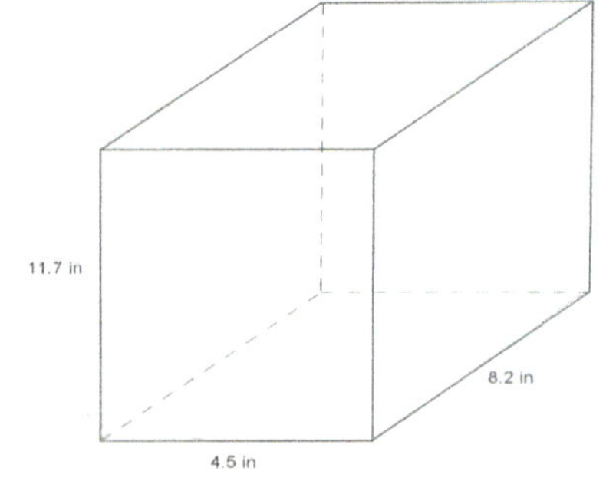

   b. What is the height?

   The height of the rectangular prism is 11.7 *in.*

   c. What is the volume of the rectangular prism?

   The volume of the rectangular prism is 431.73 *in*^{3}.

2. Use the diagram to the right to answer the questions.

   a. What is the area of the base?

   $$A = \pi 2^2$$
   $$A = 4\pi$$

   The area of the base is 4π *cm*^{2}.

   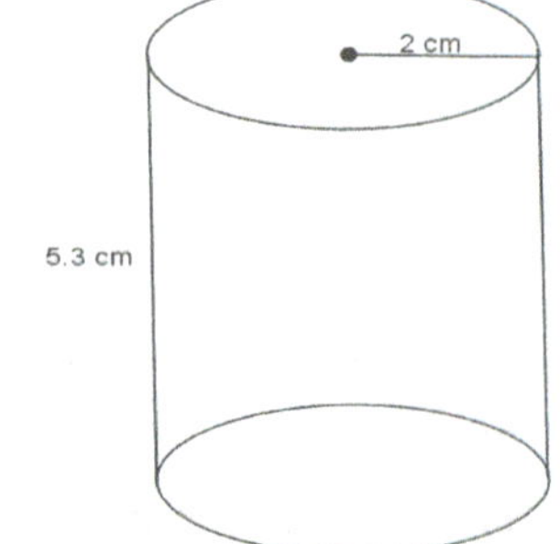

   b. What is the height?

   The height of the right circular cylinder is 5.3 *cm.*

   c. What is the volume of the right circular cylinder?

   $$V = (\pi r^2)h$$
   $$V = (4\pi)5.3$$
   $$V = 21.2\pi$$

   The volume of the right circular cylinder is 21.2π *cm*^{3}.

3. Use the diagram to the right to answer the questions.

   a. What is the area of the base?

   $$A = \pi 6^2$$
   $$A = 36\pi$$

   The area of the base is 36π *in*^{2}.

   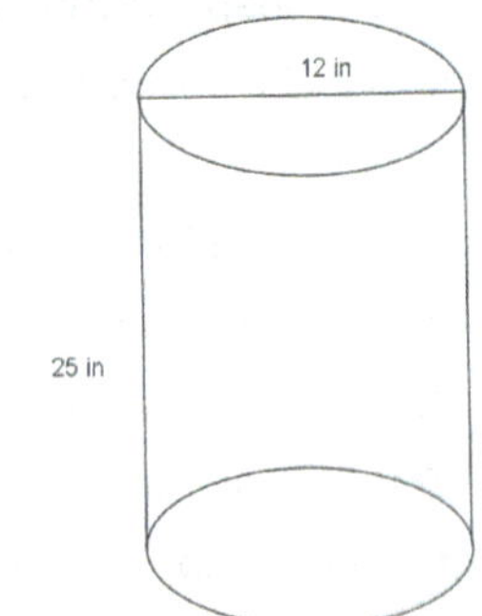

   b. What is the height?

   The height of the right circular cylinder is 25 *in.*

   c. What is the volume of the right circular cylinder?

   $$V = (36\pi)25$$
   $$V = 900\pi$$

   The volume of the right circular cylinder is 900π *in*^{3}.

      Lesson 10:   Volumes of Familiar Solids—Cones and Cylinders

EUREKA MATH™

This work is derived from Eureka Math ™ and licensed by Great Minds. ©2015 Great Minds. eureka-math.org
G8-M5-TE-B4-1.3.0-07.2015

Discussion (10 minutes)

- Next, we introduce the concept of a cone. We start with the general concept of a cylinder. Let P be a point in the plane that contains the top of a cylinder or height, h. Then, the totality of all the segments joining P to a point on the base B is a solid, called a cone, with base B and height h. The point P is the top vertex of the cone. Here are two examples of such cones.

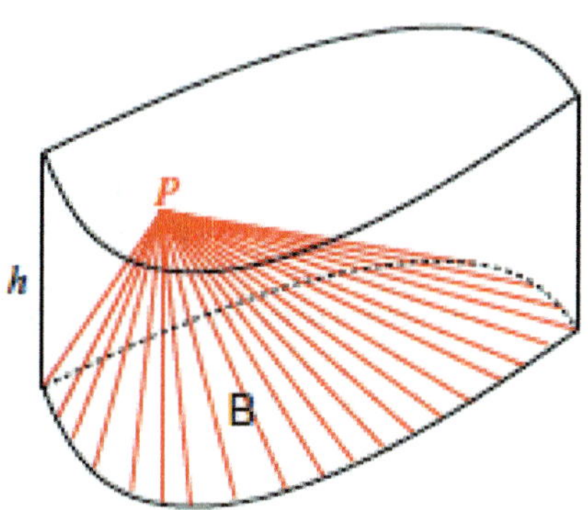

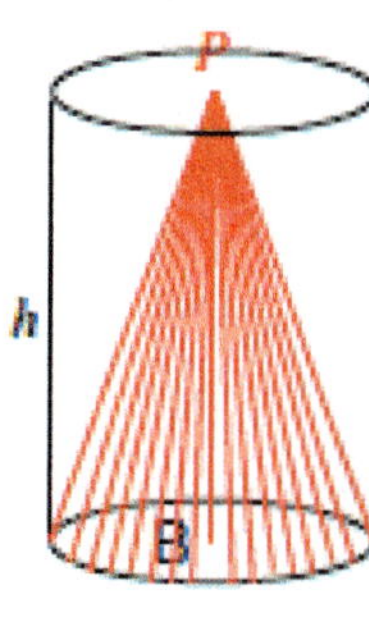

- Let's examine the diagram on the right more closely. What is the shape of the base?
    - *It appears to be the shape of a circle.*
- Where does the line segment from the vertex to the base appear to intersect the base?
    - *It appears to intersect at the center of the circle.*
- What type of angle do the line segment and base appear to make?
    - *It appears to be a right angle.*
- If the vertex of a circular cone happens to lie on the line perpendicular to the circular base at its center, then the cone is called a *right circular cone*.
- We want to develop a general formula for volume of right circular cones from our general formula for cylinders.
- If we were to fill a cone of height h and radius r with rice (or sand or water), how many cones do you think it would take to fill up a cylinder of the same height, h, and radius, r?

MP.3

Show students a cone filled with rice (or sand or water). Show students a cylinder of the same height and radius. Give students time to make a conjecture about how many cones it will take to fill the cylinder. Ask students to share their guesses and their reasoning to justify their claims. Consider having the class vote on the correct answer before the demonstration or showing the video. Demonstrate that it would take the volume of three cones to fill up the cylinder, or show the following short, one-minute video http://youtu.be/0ZACAU4SGyM.

- What would be the general formula for the volume of a right cone? Explain.

Provide students time to work in pairs to develop the formula for the volume of a cone.

- *Since it took three cones to fill up a cylinder with the same dimensions, then the volume of the cone is one-third that of the cylinder. We know the volume for a cylinder already, so the cone's volume will be $\frac{1}{3}$ of the volume of a cylinder with the same base and same height. Therefore, the formula will be $V = \frac{1}{3}(\pi r^2)h$.*

Lesson 10:   Volumes of Familiar Solids—Cones and Cylinders   **135**

This work is derived from Eureka Math ™ and licensed by Great Minds. ©2015 Great Minds. eureka-math.org
G8-M5-TE-B4-1.3.0-07.2015

Exercises 4–6 (5 minutes)

Students work independently or in pairs to complete Exercises 4–6 using the general formula for the volume of a cone. Exercise 6 is a challenge problem.

Exercises 4–6

4. Use the diagram to find the volume of the right circular cone.

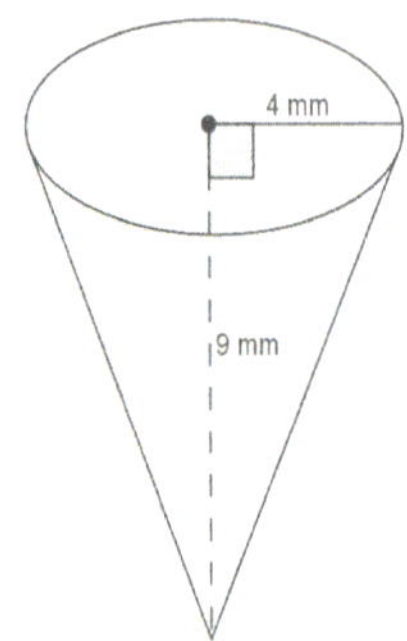

$$V = \frac{1}{3}(\pi r^2)h$$

$$V = \frac{1}{3}(\pi 4^2)9$$

$$V = 48\pi$$

The volume of the right circular cone is 48π mm^3.

5. Use the diagram to find the volume of the right circular cone.

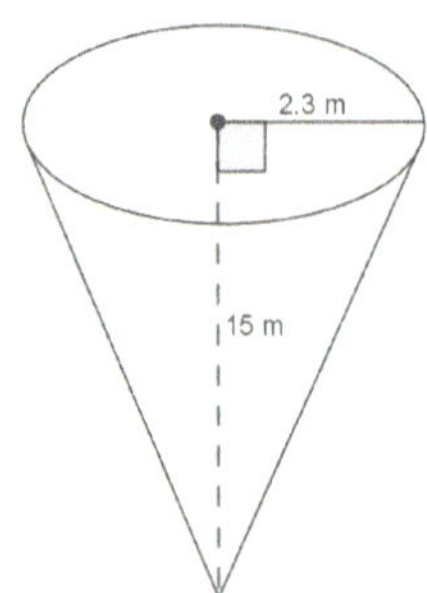

$$V = \frac{1}{3}(\pi r^2)h$$

$$V = \frac{1}{3}(\pi 2.3^2)15$$

$$V = 26.45\pi$$

The volume of the right circular cone is 26.45π m^3.

6. Challenge: A container in the shape of a right circular cone has height h, and base of radius r, as shown. It is filled with water (in its upright position) to half the height. Assume that the surface of the water is parallel to the base of the inverted cone. Use the diagram to answer the following questions:

   a. What do we know about the lengths of AB and AO?

      Then we know that $|AB| = r$, and $|AO| = h$.

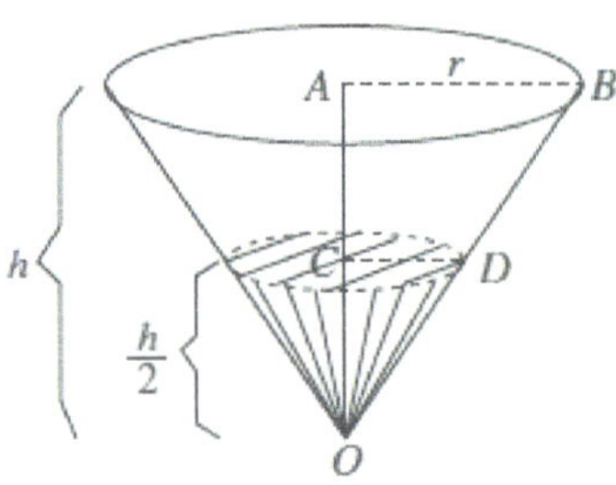

   b. What do we know about the measure of $\angle OAB$ and $\angle OCD$?

      $\angle OAB$ and $\angle OCD$ are both right angles.

   c. What can you say about $\triangle OAB$ and $\triangle OCD$?

      We have two similar triangles, $\triangle OAB$ and $\triangle OCD$ by AA criterion.

Lesson 10:   Volumes of Familiar Solids—Cones and Cylinders

EUREKA MATH

This work is derived from Eureka Math ™ and licensed by Great Minds. ©2015 Great Minds. eureka-math.org
G8-M5-TE-B4-1.3.0-07.2015

d. **What is the ratio of the volume of water to the volume of the container itself?**

Since $\dfrac{|AB|}{|CD|} = \dfrac{|AO|}{|CO|}$, and $|AO| = 2|OC|$, we have $\dfrac{|AB|}{|CD|} = \dfrac{2|OC|}{|CO|}$.

Then $|AB| = 2|CD|$.

Using the volume formula to determine the volume of the container, we have $V = \dfrac{1}{3}\pi|AB|^2|AO|$.

By substituting $|AB|$ with $2|CD|$ and $|AO|$ with $2|OC|$ we get:

$$V = \dfrac{1}{3}\pi(2|CD|)^2(2|OC|)$$

$$V = 8\left(\dfrac{1}{3}\pi|CD|^2|OC|\right),$$ *where $\dfrac{1}{3}\pi|CD|^2|OC|$ gives the volume of the portion of the container that is filled with water.*

Therefore, the volume of the water to the volume of the container is $1:8$.

Closing (4 minutes)

Summarize, or ask students to summarize, the main points from the lesson:

- Students know the volume formulas for right circular cylinders.
- Students know the volume formula for right circular cones with relation to right circular cylinders.
- Students can apply the formulas for volume of right circular cylinders and cones.

Lesson Summary

The formula to find the volume, V, of a right circular cylinder is $V = \pi r^2 h = Bh$, where B is the area of the base.

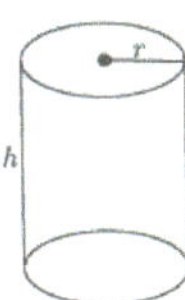

The formula to find the volume of a cone is directly related to that of the cylinder. Given a right circular cylinder with radius r and height h, the volume of a cone with those same dimensions is one-third of the cylinder. The formula for the volume, V, of a circular cone is $V = \dfrac{1}{3}\pi r^2 h$. More generally, the volume formula for a general cone is $V = \dfrac{1}{3}Bh$, where B is the area of the base.

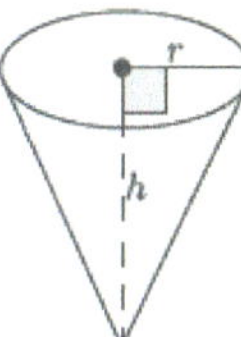

Exit Ticket (5 minutes)

This work is derived from Eureka Math ™ and licensed by Great Minds. ©2015 Great Minds. eureka-math.org
G8-M5-TE-B4-1.3.0-07.2015

Name ____   Date _______________________

Lesson 10: Volumes of Familiar Solids—Cones and Cylinders

Exit Ticket

1. Use the diagram to find the total volume of the three cones shown below.

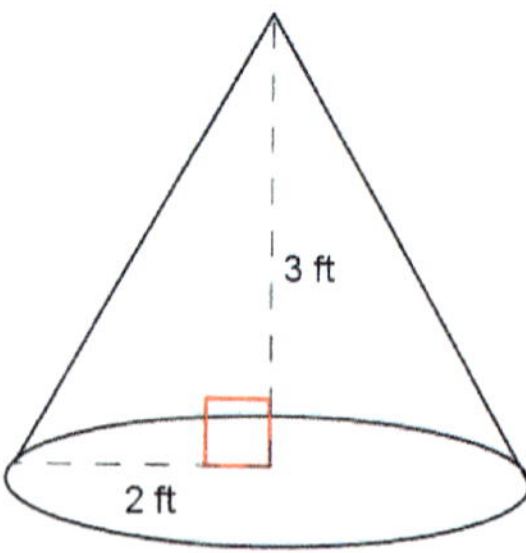

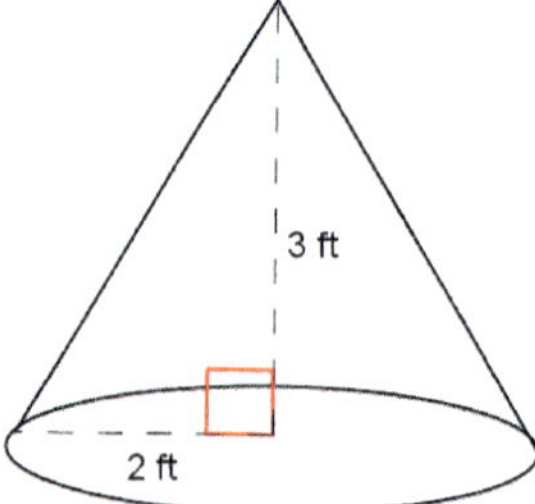

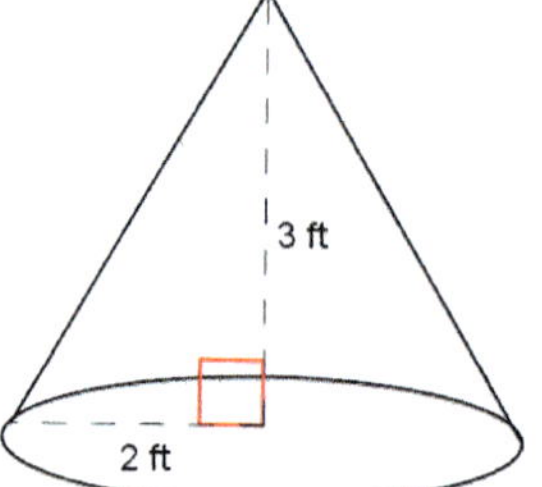

2. Use the diagram below to determine which has the greater volume, the cone or the cylinder.

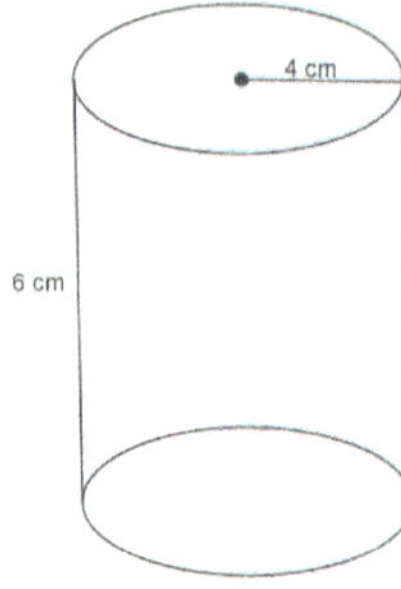

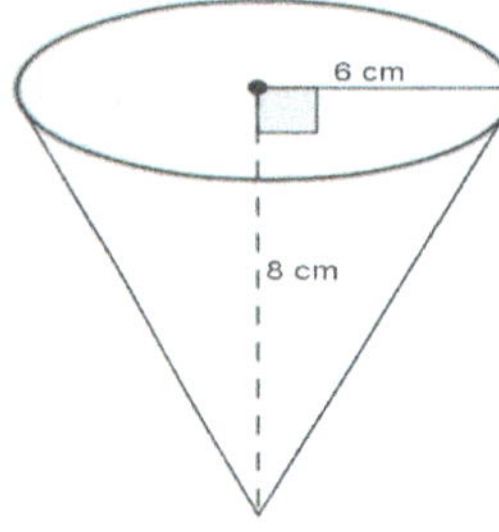

   Lesson 10:   Volumes of Familiar Solids—Cones and Cylinders

EUREKA MATH

This work is derived from Eureka Math ™ and licensed by Great Minds. ©2015 Great Minds. eureka-math.org
G8-M5-TE-B4-1.3.0-07.2015

Exit Ticket Sample Solutions

1. **Use the diagram to find the total volume of the three cones shown below.**

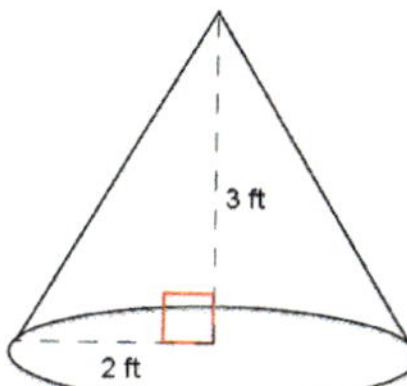 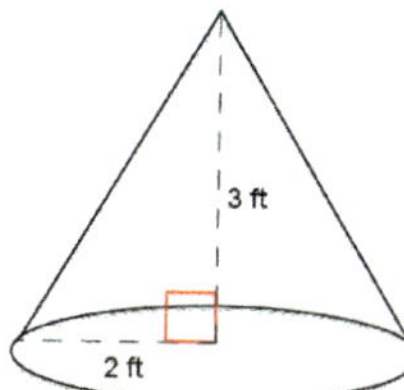 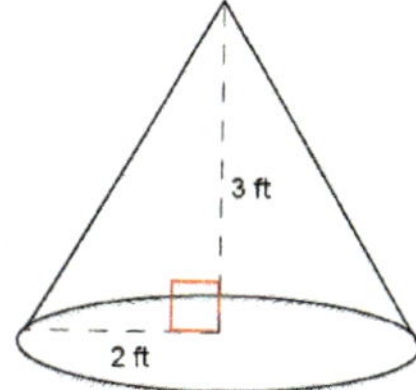

Since all three cones have the same base and height, the volume of the three cones will be the same as finding the volume of a cylinder with the same base radius and same height.

$$V = \pi r^2 h$$
$$V = \pi (2)^2 3$$
$$V = 12\pi$$

The volume of all three cones is 12π ft^3.

2. **Use the diagram below to determine which has the greater volume, the cone or the cylinder.**

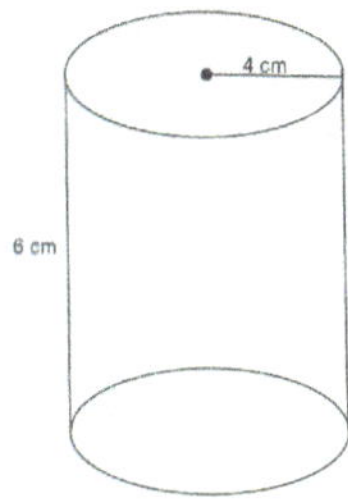 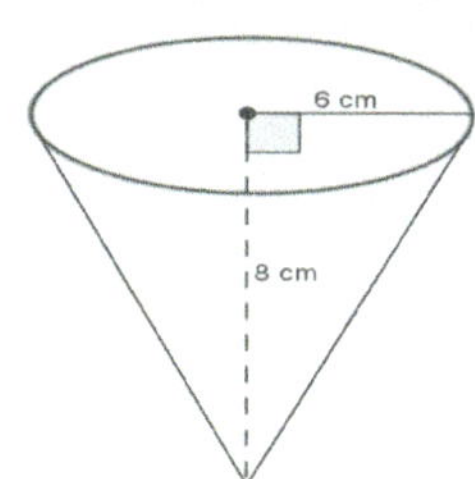

$$V = \pi r^2 h$$
$$V = \pi 4^2 (6)$$
$$V = 96\pi$$

The volume of the cylinder is 96π cm^3.

$$V = \frac{1}{3}\pi r^2 h$$
$$V = \frac{1}{3}\pi 6^2 (8)$$
$$V = 96\pi$$

The volume of the cone is 96π cm^3.

The volume of the cylinder and the volume of the cone are the same, 96π cm^3.

This work is derived from Eureka Math ™ and licensed by Great Minds. ©2015 Great Minds. eureka-math.org
G8-M5-TE-B4-1.3.0-07.2015

Problem Set Sample Solutions

1. Use the diagram to help you find the volume of the right circular cylinder.

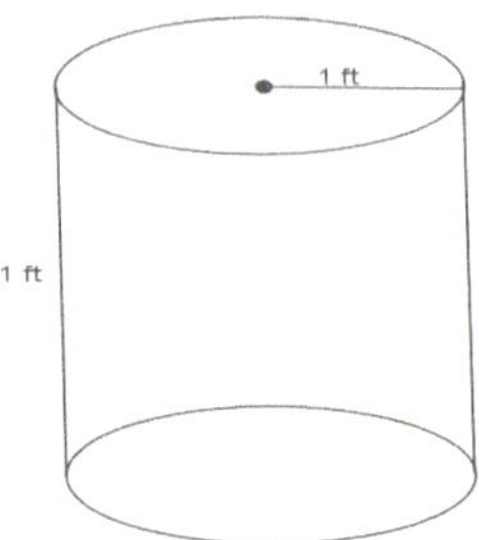

$$V = \pi r^2 h$$
$$V = \pi (1)^2 (1)$$
$$V = \pi$$

The volume of the right circular cylinder is π ft^3.

2. Use the diagram to help you find the volume of the right circular cone.

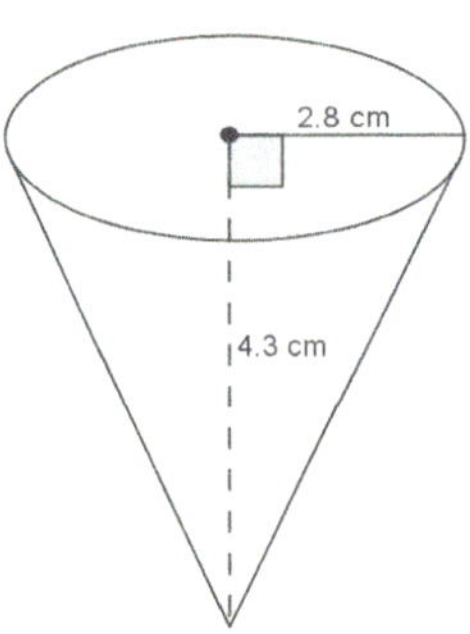

$$V = \frac{1}{3} \pi r^2 h$$
$$V = \frac{1}{3} \pi (2.8)^2 (4.3)$$
$$V = 11.237333\ldots \pi$$

The volume of the right circular cone is about 11.2π cm^3.

    Lesson 10:   Volumes of Familiar Solids—Cones and Cylinders

EUREKA MATH™

This work is derived from Eureka Math ™ and licensed by Great Minds. ©2015 Great Minds. eureka-math.org
G8-M5-TE-B4-1.3.0-07.2015

3. Use the diagram to help you find the volume of the right circular cylinder.

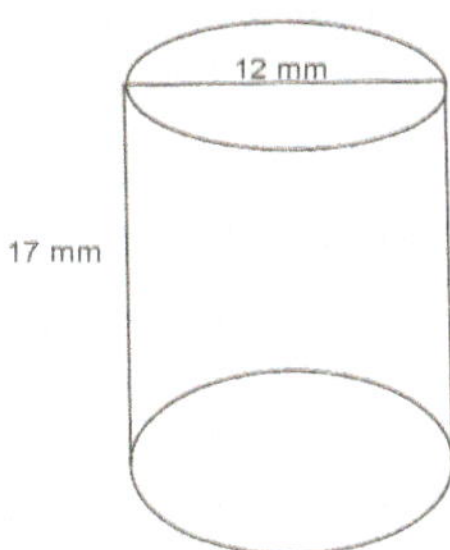

If the diameter is 12 mm, then the radius is 6 mm.

$$V = \pi r^2 h$$
$$V = \pi (6)^2 (17)$$
$$V = 612\pi$$

The volume of the right circular cylinder is 612π mm³.

4. Use the diagram to help you find the volume of the right circular cone.

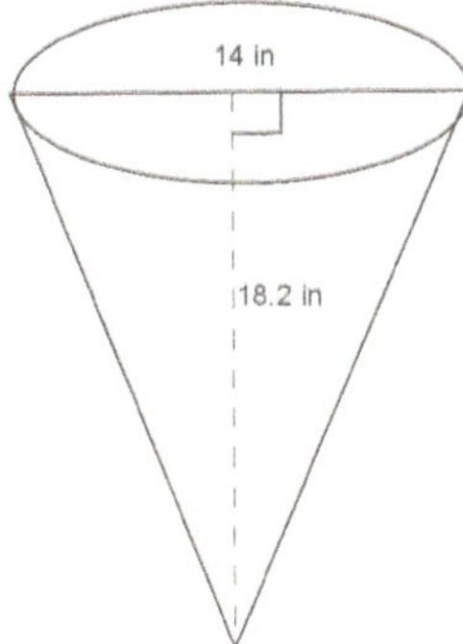

If the diameter is 14 in., then the radius is 7 in.

$$V = \frac{1}{3}\pi r^2 h$$
$$V = \frac{1}{3}\pi (7)^2 (18.2)$$
$$V = 297.26666\ldots\pi$$
$$V \approx 297.3\pi$$

The volume of the right cone is about 297.3π in³.

This work is derived from Eureka Math ™ and licensed by Great Minds. ©2015 Great Minds. eureka-math.org
G8-M5-TE-B4-1.3.0-07.2015

5. Oscar wants to fill with water a bucket that is the shape of a right circular cylinder. It has a 6-inch radius and 12-inch height. He uses a shovel that has the shape of a right circular cone with a 3-inch radius and 4-inch height. How many shovelfuls will it take Oscar to fill the bucket up level with the top?

$$V = \pi r^2 h$$
$$V = \pi(6)^2(12)$$
$$V = 432\pi$$

The volume of the bucket is 432π in^3.

$$V = \frac{1}{3}\pi r^2 h$$
$$V = \frac{1}{3}\pi(3)^2(4)$$
$$V = 12\pi$$

The volume of shovel is 12π in^3.

$$\frac{432\pi}{12\pi} = 36$$

It would take 36 shovelfuls of water to fill up the bucket.

6. A cylindrical tank (with dimensions shown below) contains water that is 1-foot deep. If water is poured into the tank at a constant rate of $20 \ \frac{\text{ft}^3}{\text{min}}$ for 20 min., will the tank overflow? Use 3.14 to estimate π.

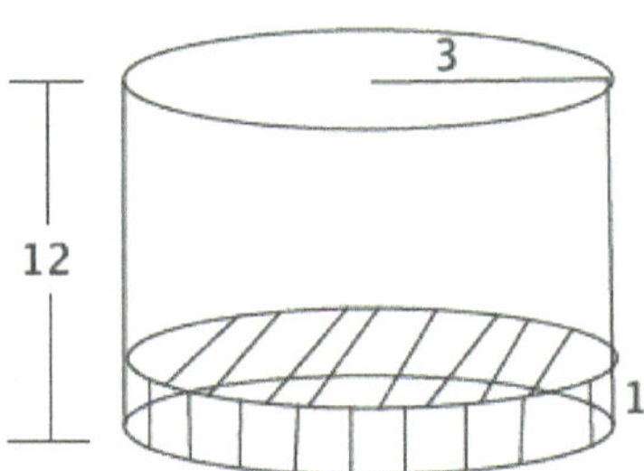

$$V = \pi r^2 h$$
$$V = \pi(3)^2(12)$$
$$V = 108\pi$$

The volume of the tank is about 339.12 ft³.

$$V = \pi r^2 h$$
$$V = \pi(3)^2(1)$$
$$V = 9\pi$$

There is about 28.26 ft³ of water already in the tank. There is about 310.86 ft³ of space left in the tank. If the water is poured at a constant rate for 20 min., 400 ft³ will be poured into the tank, and the tank will overflow.

Lesson 10:   Volumes of Familiar Solids—Cones and Cylinders

EUREKA MATH™

This work is derived from Eureka Math ™ and licensed by Great Minds. ©2015 Great Minds. eureka-math.org
G8-M5-TE-B4-1.3.0-07.2015

Lesson 11: Volume of a Sphere

Student Outcomes

- Students know the volume formula for a sphere as it relates to a right circular cylinder with the same diameter and height.
- Students apply the formula for the volume of a sphere to real-world and mathematical problems.

Lesson Notes

The demonstrations in this lesson require a sphere (preferably one that can be filled with water, sand, or rice) and a right circular cylinder with the same diameter and height as the diameter of the sphere. Demonstrate to students that the volume of a sphere is two-thirds the volume of the circumscribing cylinder. If this demonstration is impossible, a video link is included to show a demonstration.

Classwork

Discussion (10 minutes)

Show students pictures of the spheres shown below (or use real objects). Ask the class to come up with a mathematical definition on their own.

- Finally, we come to the volume of a sphere of radius r. First recall that a sphere of radius r is the set of all the points in three-dimensional space of distance r from a fixed point, called the center of the sphere. So a sphere is, by definition, a surface, or a two-dimensional object. When we talk about the volume of a sphere, we mean the volume of the solid inside this surface.

> **Scaffolding:**
> Consider using a small bit of clay to represent the center and toothpicks to represent the radius of a sphere.

- The discovery of this formula was a major event in ancient mathematics. The first person to discover the formula was Archimedes (287–212 B.C.E.), but it was also independently discovered in China by Zu Chongshi (429–501 C.E.) and his son Zu Geng (circa 450–520 C.E.) by essentially the same method. This method has come to be known as *Cavalieri's Principle* because he announced this method at a time when he had an audience. Cavalieri (1598–1647) was one of the forerunners of calculus.

Lesson 11:   Volume of a Sphere   143

This work is derived from Eureka Math ™ and licensed by Great Minds. ©2015 Great Minds. eureka-math.org
G8-M5-TE-B4-1.3.0-07.2015

Show students a cylinder. Convince them that the diameter of the sphere is the same as the diameter and the height of the cylinder. Give students time to make a conjecture about how much of the volume of the cylinder is taken up by the sphere. Ask students to share their guesses and their reasoning. Consider having the class vote on the correct answer before proceeding with the discussion.

- The derivation of this formula and its understanding requires advanced mathematics, so we will not derive it at this time.

If possible, do a physical demonstration showing that the volume of a sphere is exactly $\frac{2}{3}$ the volume of a cylinder with the same diameter and height. Also consider showing the following 1:17-minute video: http://www.youtube.com/watch?v=aLyQddyY8ik.

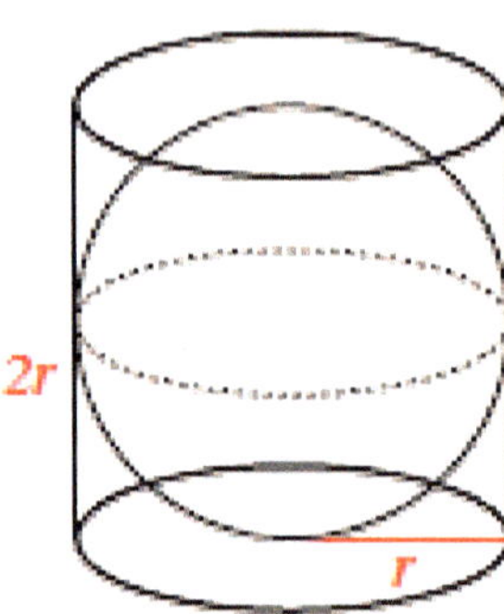

- Based on the demonstration (or video), we can say that

$$\text{Volume(sphere)} = \frac{2}{3}\,\text{volume(cylinder with same diameter and height of the sphere)}.$$

Exercises 1–3 (5 minutes)

Students work independently or in pairs using the general formula for the volume of a sphere. Verify that students are able to compute the formula for the volume of a sphere.

Exercises 1–3

1. **What is the volume of a cylinder?**

   $V = \pi r^2 h$

2. **What is the height of the cylinder?**

   The height of the cylinder is the same as the diameter of the sphere. The diameter is $2r$.

MP.2

3. **If volume(sphere) $= \frac{2}{3}$ volume(cylinder with same diameter and height), what is the formula for the volume of a sphere?**

$$\text{Volume(sphere)} = \frac{2}{3}(\pi r^2 h)$$

$$\text{Volume(sphere)} = \frac{2}{3}(\pi r^2 2r)$$

$$\text{Volume(sphere)} = \frac{4}{3}(\pi r^3)$$

Lesson 11:   Volume of a Sphere

EUREKA MATH™

This work is derived from Eureka Math ™ and licensed by Great Minds. ©2015 Great Minds. eureka-math.org
G8-M5-TE-B4-1.3.0-07.2015

Example 1 (4 minutes)

- When working with circular two- and three-dimensional figures, we can express our answers in two ways. One is exact and will contain the symbol for pi, π. The other is an approximation, which usually uses 3.14 for π. Unless noted otherwise, we will have exact answers that contain the pi symbol.

- For Examples 1 and 2, use the formula from Exercise 3 to compute the exact volume for the sphere shown below.

Example 1

Compute the exact volume for the sphere shown below.

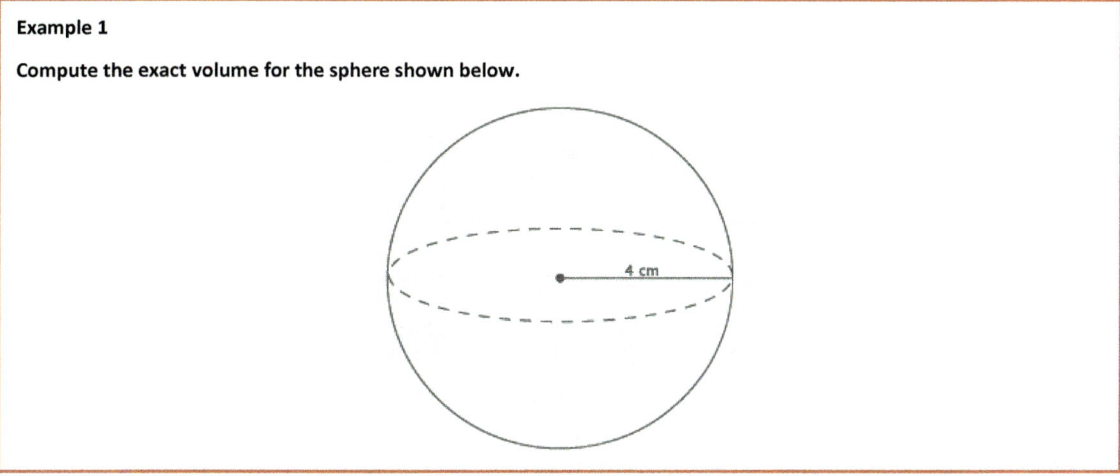

Provide students time to work; then, have them share their solutions.

- *Sample student work:*

$$V = \frac{4}{3}\pi r^3$$

$$= \frac{4}{3}\pi (4^3)$$

$$= \frac{4}{3}\pi (64)$$

$$= \frac{256}{3}\pi$$

$$= 85\frac{1}{3}\pi$$

The volume of the sphere is $85\frac{1}{3}\pi$ cm^3.

EUREKA MATH™

This work is derived from Eureka Math ™ and licensed by Great Minds. ©2015 Great Minds. eureka-math.org
G8-M5-TE-B4-1.3.0-07.2015

Example 2 (6 minutes)

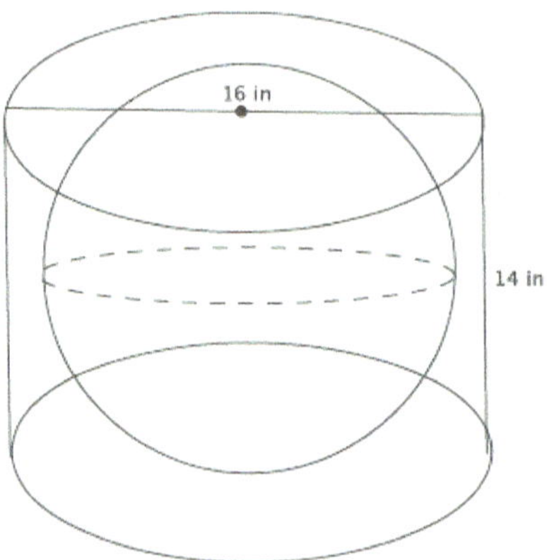

> **Example 2**
>
> A cylinder has a diameter of 16 inches and a height of 14 inches. What is the volume of the largest sphere that will fit into the cylinder?
>
>

- What is the radius of the base of the cylinder?
  - *The radius of the base of the cylinder is 8 inches.*
- Could the sphere have a radius of 8 inches? Explain.
  - *No. If the sphere had a radius of 8 inches, then it would not fit into the cylinder because the height is only 14 inches. With a radius of 8 inches, the sphere would have a height of $2r$, or 16 inches. Since the cylinder is only 14 inches high, the radius of the sphere cannot be 8 inches.*
- What size radius for the sphere would fit into the cylinder? Explain.
  - *A radius of 7 inches would fit into the cylinder because $2r$ is 14, which means the sphere would touch the top and bottom of the cylinder. A radius of 7 means the radius of the sphere would not touch the sides of the cylinder, but would fit into it.*
- Now that we know the radius of the largest sphere is 7 inches, what is the volume of the sphere?
  - *Sample student work:*

$$V = \frac{4}{3}\pi r^3$$

$$= \frac{4}{3}\pi(7^3)$$

$$= \frac{4}{3}\pi(343)$$

$$= \frac{1372}{3}\pi$$

$$= 457\frac{1}{3}\pi$$

The volume of the sphere is $457\frac{1}{3}\pi$ cm^3.

Lesson 11:   Volume of a Sphere

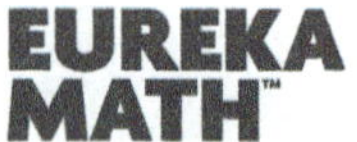

This work is derived from Eureka Math ™ and licensed by Great Minds. ©2015 Great Minds. eureka-math.org
G8-M5-TE-B4-1.3.0-07.2015

Exercises 4–8 (10 minutes)

Students work independently or in pairs to use the general formula for the volume of a sphere.

Exercises 4–8

4. Use the diagram and the general formula to find the volume of the sphere.

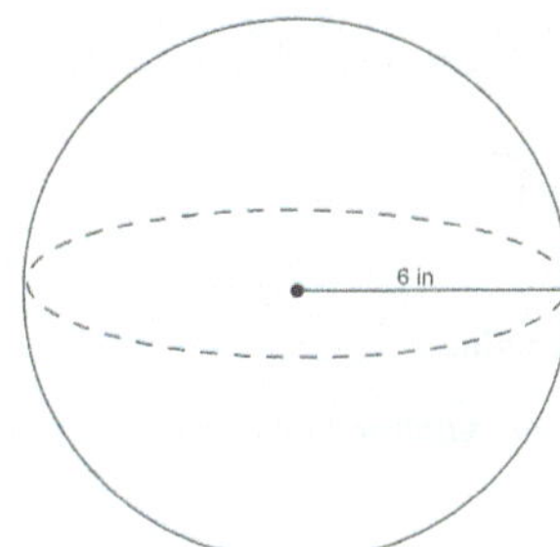

$$V = \frac{4}{3}\pi r^3$$

$$V = \frac{4}{3}\pi(6^3)$$

$$V \approx 288\pi$$

The volume of the sphere is about 288π in^3.

5. The average basketball has a diameter of 9.5 inches. What is the volume of an average basketball? Round your answer to the tenths place.

$$V = \frac{4}{3}\pi r^3$$

$$V = \frac{4}{3}\pi(4.75^3)$$

$$V = \frac{4}{3}\pi(107.17)$$

$$V \approx 142.9\pi$$

The volume of an average basketball is about 142.9π in^3.

6. A spherical fish tank has a radius of 8 inches. Assuming the entire tank could be filled with water, what would the volume of the tank be? Round your answer to the tenths place.

$$V = \frac{4}{3}\pi r^3$$

$$V = \frac{4}{3}\pi(8^3)$$

$$V = \frac{4}{3}\pi(512)$$

$$V \approx 682.7\pi$$

The volume of the fish tank is about 682.7π in^3.

This work is derived from Eureka Math ™ and licensed by Great Minds. ©2015 Great Minds. eureka-math.org
G8-M5-TE-B4-1.3.0-07.2015

7. Use the diagram to answer the questions.

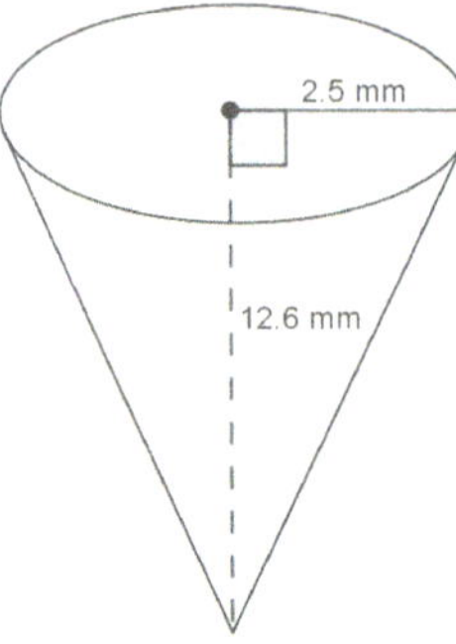

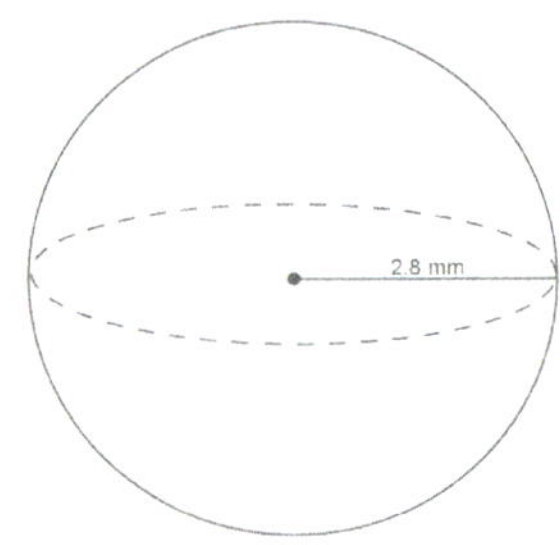

a. Predict which of the figures shown above has the greater volume. Explain.

Student answers will vary. Students will probably say the cone has more volume because it looks larger.

b. Use the diagram to find the volume of each, and determine which has the greater volume.

$$V = \frac{1}{3}\pi r^2 h$$

$$V = \frac{1}{3}\pi (2.5^2)(12.6)$$

$$V = 26.25\pi$$

The volume of the cone is 26.25π mm^3.

$$V = \frac{4}{3}\pi r^3$$

$$V = \frac{4}{3}\pi (2.8^3)$$

$$V = 29.269333 \ldots \pi$$

The volume of the sphere is about 29.27π mm^3. The volume of the sphere is greater than the volume of the cone.

8. One of two half spheres formed by a plane through the sphere's center is called a hemisphere. What is the formula for the volume of a hemisphere?

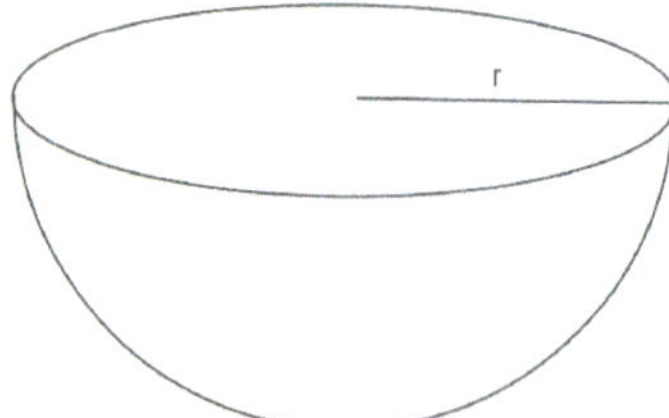

Since a hemisphere is half a sphere, the volume(hemisphere) $= \frac{1}{2}$ (volume of sphere).

$$V = \frac{1}{2}\left(\frac{4}{3}\pi r^3\right)$$

$$V = \frac{2}{3}\pi r^3$$

Lesson 11:   Volume of a Sphere

EUREKA
MATH

This work is derived from Eureka Math ™ and licensed by Great Minds. ©2015 Great Minds. eureka-math.org
G8-M5-TE-B4-1.3.0-07.2015

Closing (5 minutes)

Summarize, or ask students to summarize, the main points from the lesson:

- Students know the volume formula for a sphere with relation to a right circular cylinder.
- Students know the volume formula for a hemisphere.
- Students can apply the volume of a sphere to solve mathematical problems.

Lesson Summary

The formula to find the volume of a sphere is directly related to that of the right circular cylinder. Given a right circular cylinder with radius r and height h, which is equal to $2r$, a sphere with the same radius r has a volume that is exactly two-thirds of the cylinder.

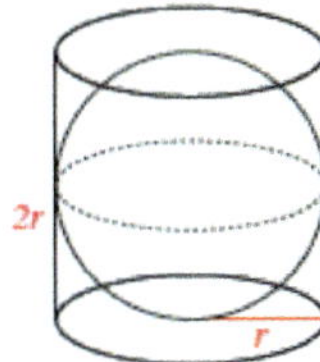

Therefore, the volume of a sphere with radius r has a volume given by the formula $V = \frac{4}{3}\pi r^3$.

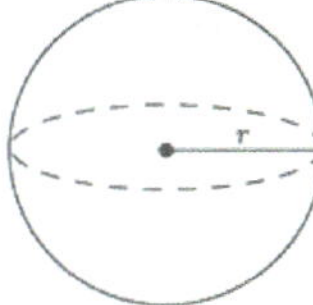

Exit Ticket (5 minutes)

This work is derived from Eureka Math ™ and licensed by Great Minds. ©2015 Great Minds. eureka-math.org
G8-M5-TE-B4-1.3.0-07.2015

Name _______________________________________   Date _______________________

Lesson 11: Volume of a Sphere

Exit Ticket

1. What is the volume of the sphere shown below?

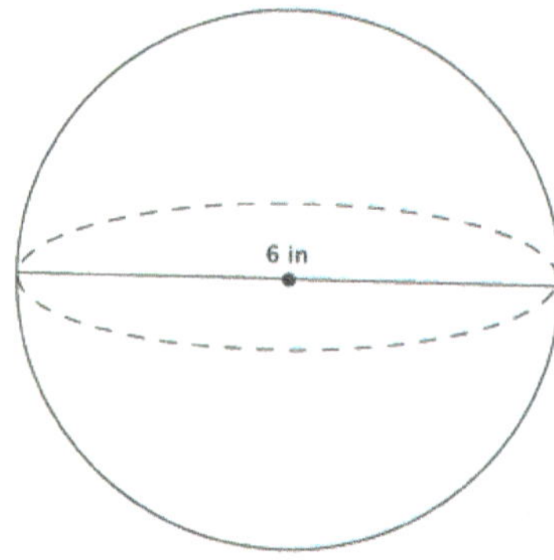

2. Which of the two figures below has the greater volume?

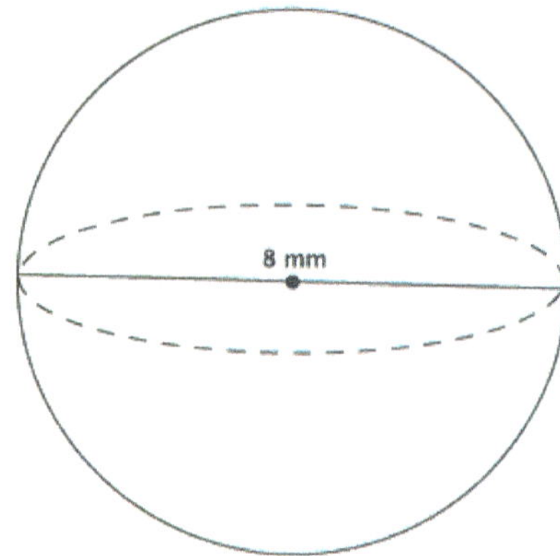

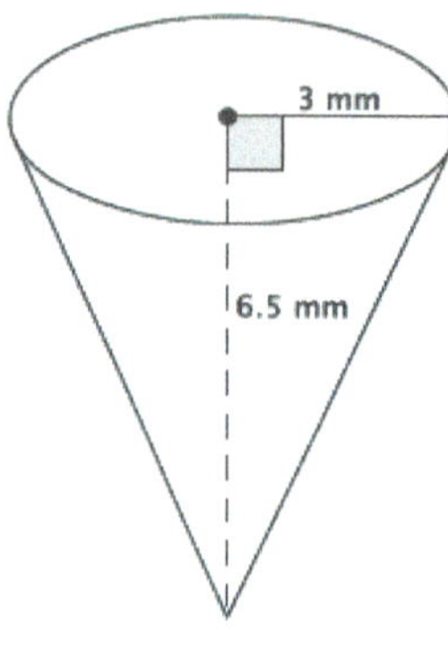

　　Lesson 11:　　Volume of a Sphere

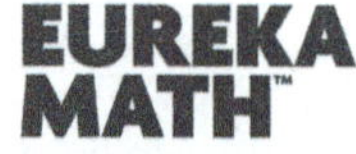

This work is derived from Eureka Math ™ and licensed by Great Minds. ©2015 Great Minds. eureka-math.org
G8-M5-TE-B4-1.3.0-07.2015

Exit Ticket Sample Solutions

1. What is the volume of the sphere shown below?

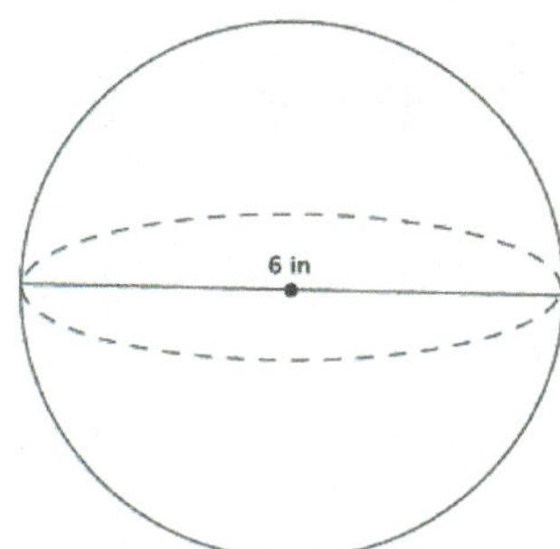

$$V = \frac{4}{3}\pi r^3$$

$$= \frac{4}{3}\pi (3^3)$$

$$= \frac{108}{3}\pi$$

$$= 36\pi$$

The volume of the sphere is 36π in^3.

2. Which of the two figures below has the greater volume?

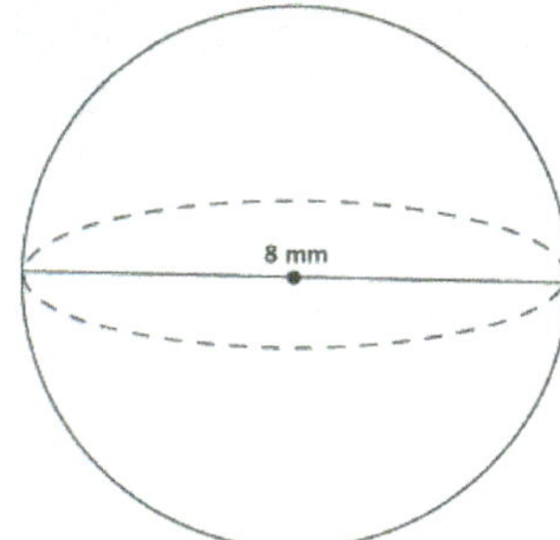

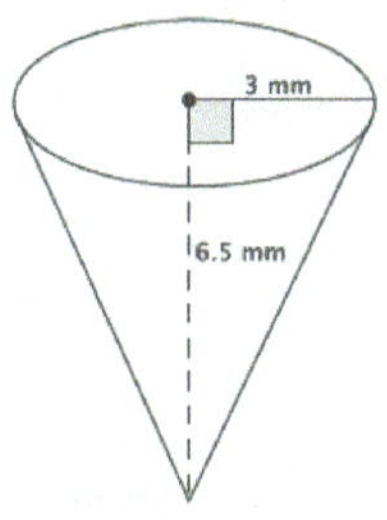

$$V = \frac{4}{3}\pi r^3$$

$$= \frac{4}{3}\pi (4^3)$$

$$= \frac{256}{3}\pi$$

$$= 85\frac{1}{3}\pi$$

The volume of the sphere is $85\frac{1}{3}\pi$ mm^3.

$$V = \frac{1}{3}\pi r^2 h$$

$$= \frac{1}{3}\pi (3^2)(6.5)$$

$$= \frac{58.5}{3}\pi$$

$$= 19.5\pi$$

The volume of the cone is 19.5π mm^3. The sphere has the greater volume.

EUREKA MATH™

This work is derived from Eureka Math ™ and licensed by Great Minds. ©2015 Great Minds. eureka-math.org
G8-M5-TE-B4-1.3.0-07.2015

Problem Set Sample Solutions

1. Use the diagram to find the volume of the sphere.

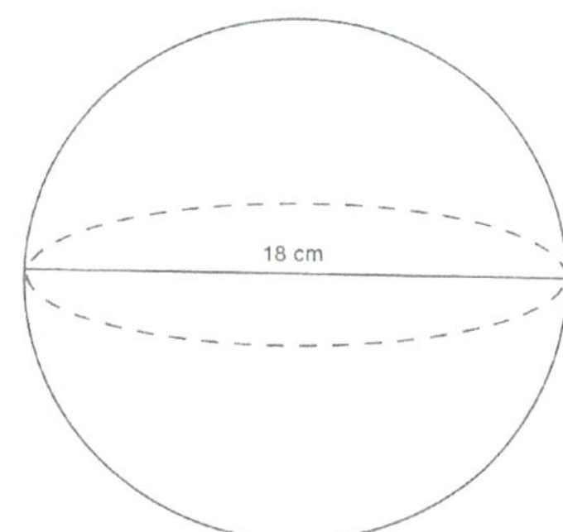

$$V = \frac{4}{3}\pi r^3$$

$$V = \frac{4}{3}\pi (9^3)$$

$$V = 972\pi$$

The volume of the sphere is 972π cm^3.

2. Determine the volume of a sphere with diameter 9 mm, shown below.

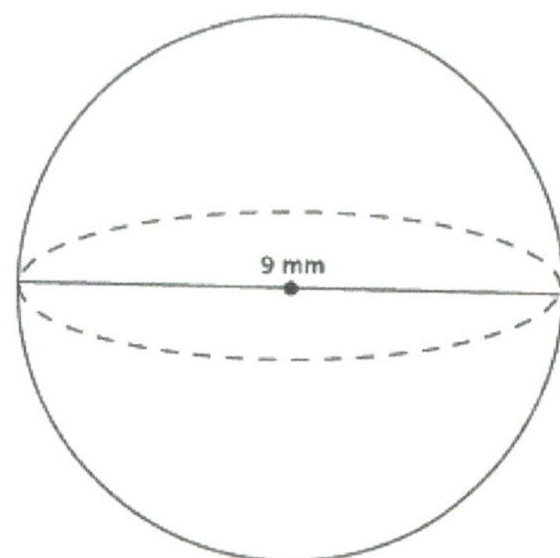

$$V = \frac{4}{3}\pi r^3$$

$$= \frac{4}{3}\pi (4.5^3)$$

$$= \frac{364.5}{3}\pi$$

$$= 121.5\pi$$

The volume of the sphere is 121.5π mm^3.

3. Determine the volume of a sphere with diameter 22 in., shown below.

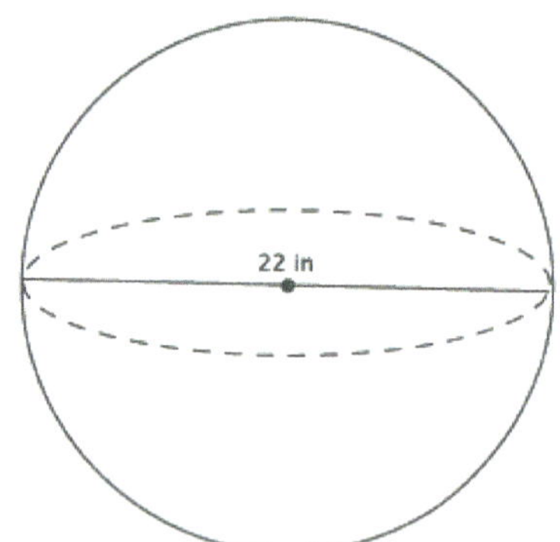

$$V = \frac{4}{3}\pi r^3$$

$$= \frac{4}{3}\pi (11^3)$$

$$= \frac{5324}{3}\pi$$

$$= 1774\frac{2}{3}\pi$$

The volume of the sphere is $1774\frac{2}{3}\pi$ in^3.

EUREKA MATH

This work is derived from Eureka Math ™ and licensed by Great Minds. ©2015 Great Minds. eureka-math.org
G8-M5-TE-B4-1.3.0-07.2015

4. Which of the two figures below has the lesser volume?

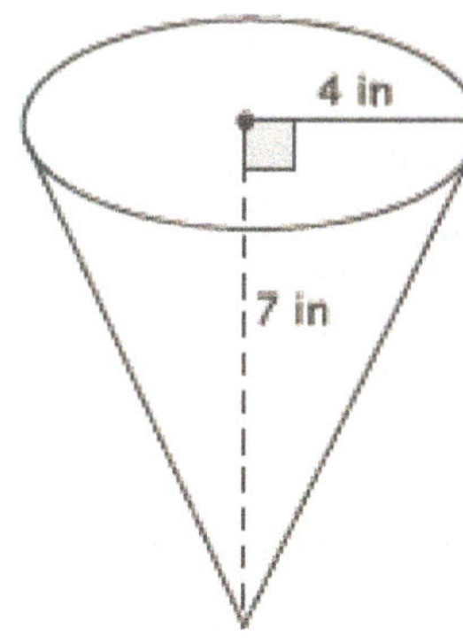

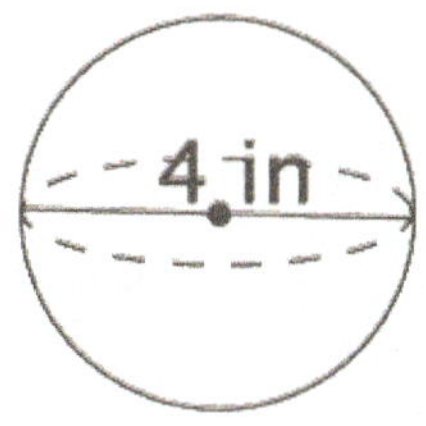

The volume of the cone:

$$V = \frac{1}{3}\pi r^2 h$$
$$= \frac{1}{3}\pi(16)(7)$$
$$= \frac{112}{3}\pi$$
$$= 37\frac{1}{3}\pi$$

The volume of the sphere:

$$V = \frac{4}{3}\pi r^3$$
$$= \frac{4}{3}\pi(2^3)$$
$$= \frac{32}{3}\pi$$
$$= 10\frac{2}{3}\pi$$

The cone has volume $37\frac{1}{3}\pi$ in³ and the sphere has volume $10\frac{2}{3}\pi$ in³. The sphere has the lesser volume.

5. Which of the two figures below has the greater volume?

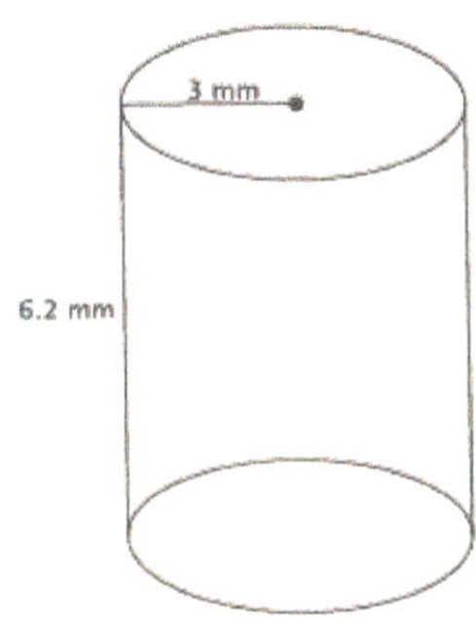

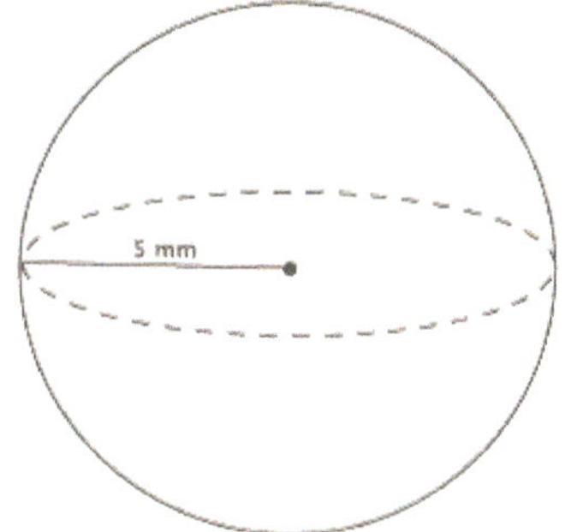

The volume of the cylinder:

$$V = \pi r^2 h$$
$$= \pi(3^2)(6.2)$$
$$= 55.8\pi$$

The volume of the sphere:

$$V = \frac{4}{3}\pi r^3$$
$$= \frac{4}{3}\pi(5^3)$$
$$= \frac{500}{3}\pi$$
$$= 166\frac{2}{3}\pi$$

The cylinder has volume 55.8π mm³ and the sphere has volume $166\frac{2}{3}\pi$ mm³. The sphere has the greater volume.

This work is derived from Eureka Math ™ and licensed by Great Minds. ©2015 Great Minds. eureka-math.org
G8-M5-TE-B4-1.3.0-07.2015

6. Bridget wants to determine which ice cream option is the best choice. The chart below gives the description and prices for her options. Use the space below each item to record your findings.

$2.00	$3.00	$4.00
One scoop in a cup	Two scoops in a cup	Three scoops in a cup
$V \approx 4.19 \text{ in}^3$	$V \approx 8.37 \text{ in}^3$	$V \approx 12.56 \text{ in}^3$
Half a scoop on a cone filled with ice cream		A cup filled with ice cream (level to the top of the cup)
$V \approx 6.8 \text{ in}^3$		$V \approx 14.13 \text{ in}^3$

A scoop of ice cream is considered a perfect sphere and has a 2-inch diameter. A cone has a 2-inch diameter and a height of 4.5 inches. A cup, considered a right circular cylinder, has a 3-inch diameter and a height of 2 inches.

a. Determine the volume of each choice. Use 3.14 to approximate π.

First, find the volume of one scoop of ice cream.

$$\text{Volume of one scoop} = \frac{4}{3}\pi(1^3)$$

The volume of one scoop of ice cream is $\frac{4}{3}\pi$ in^3, or approximately 4.19 in^3.

The volume of two scoops of ice cream is $\frac{8}{3}\pi$ in^3, or approximately 8.37 in^3.

The volume of three scoops of ice cream is $\frac{12}{3}\pi$ in^3, or approximately 12.56 in^3.

$$\text{Volume of half scoop} = \frac{2}{3}\pi(1^3)$$

The volume of half a scoop of ice cream is $\frac{2}{3}\pi$ in^3, or approximately 2.09 in^3.

$$\text{Volume of cone} = \frac{1}{3}(\pi r^2)h$$
$$V = \frac{1}{3}\pi(1^2)4.5$$
$$V = 1.5\pi$$

The volume of the cone is 1.5π in^3, or approximately 4.71 in^3. Then, the cone with half a scoop of ice cream on top is approximately 6.8 in^3.

$$V = \pi r^2 h$$
$$V = \pi 1.5^2(2)$$
$$V = 4.5\pi$$

The volume of the cup is 4.5π in^3, or approximately 14.13 in^3.

b. Determine which choice is the best value for her money. Explain your reasoning.

Student answers may vary.

Checking the cost for every in^3 of each choice:

$$\frac{2}{4.19} \approx 0.47723\ldots$$
$$\frac{2}{6.8} \approx 0.29411\ldots$$
$$\frac{3}{8.37} \approx 0.35842\ldots$$
$$\frac{4}{12.56} \approx 0.31847\ldots$$
$$\frac{4}{14.13} \approx 0.28308\ldots$$

The best value for her money is the cup filled with ice cream since it costs about 28 cents for every in^3.

Lesson 11:   Volume of a Sphere

EUREKA MATH™

This work is derived from Eureka Math ™ and licensed by Great Minds. ©2015 Great Minds. eureka-math.org
G8-M5-TE-B4-1.3.0-07.2015

Name ____   Date _________________

1.

a. We define x as a year between 2008 and 2013 and y as the total number of smartphones sold that year, in millions. The table shows values of x and corresponding y values.

Year (x)	2008	2009	2010	2011	2012	2013
Number of smartphones in millions (y)	3.7	17.3	42.4	90	125	153.2

i. How many smartphones were sold in 2009?

ii. In which year were 90 million smartphones sold?

iii. Is y a function of x? Explain why or why not.

b. Randy began completing the table below to represent a particular linear function. Write an equation to represent the function he was using and complete the table for him.

Input (x)	-3	-1	0	$\dfrac{1}{2}$	1	2	3
Output (y)	-5		4				13

This work is derived from Eureka Math ™ and licensed by Great Minds. ©2015 Great Minds. eureka-math.org
G8-M5-TE-B4-1.3.0-07.2015

c.   Create the graph of the function in part (b).

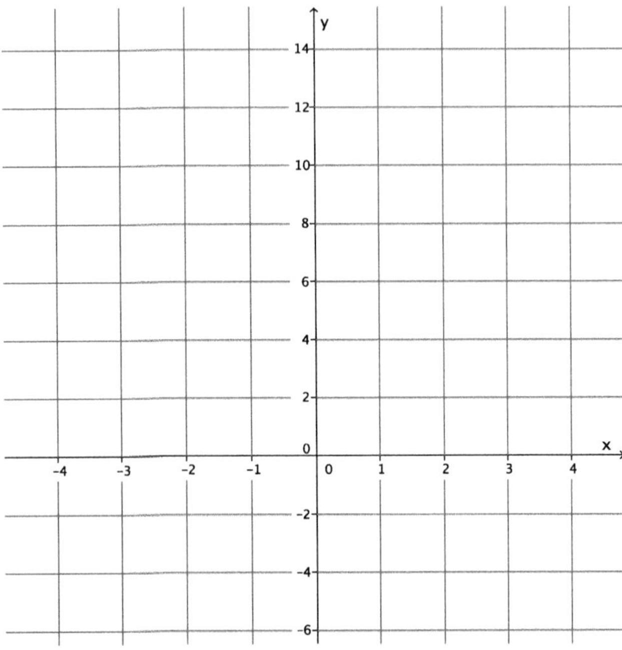

d.   At NYU in 2013, the cost of the weekly meal plan options could be described as a function of the number of meals. Is the cost of the meal plan a linear or nonlinear function? Explain.

8 meals: $125/week
10 meals: $135/week
12 meals: $155/week
21 meals: $220/week

This work is derived from Eureka Math ™ and licensed by Great Minds. ©2015 Great Minds. eureka-math.org
G8-M5-TE-B4-1.3.0-07.2015

EUREKA MATH™

2. The cost to enter and go on rides at a local water park, Wally's Water World, is shown in the graph below.

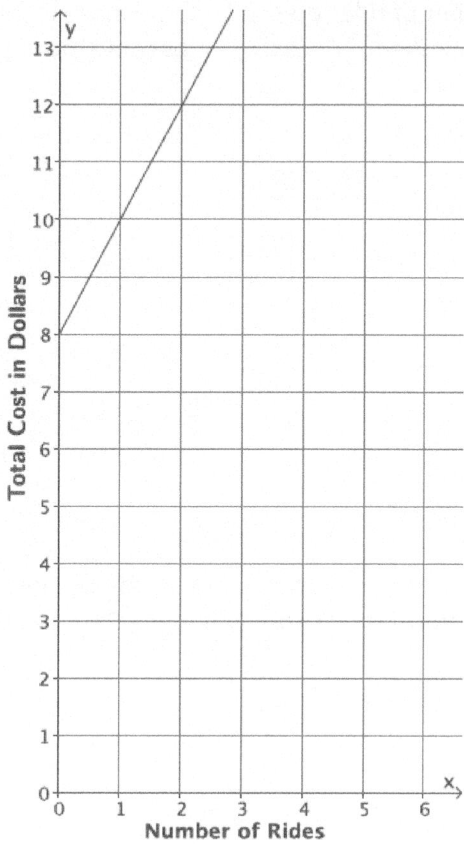

A new water park, Tony's Tidal Takeover, just opened. You have not heard anything specific about how much it costs to go to this park, but some of your friends have told you what they spent. The information is organized in the table below.

Number of rides	0	2	4	6
Dollars spent	12.00	13.50	15.00	16.50

Each park charges a different admission fee and a different fee per ride, but the cost of each ride remains the same.

a. If you only have $14 to spend, which park would you attend (assume the rides are the same quality)? Explain.

This work is derived from Eureka Math ™ and licensed by Great Minds. ©2015 Great Minds. eureka-math.org
G8-M5-TE-B4-1.3.0-07.2015

b. Another water park, Splash, opens, and they charge an admission fee of $30 with no additional fee for rides. At what number of rides does it become more expensive to go to Wally's Water World than Splash? At what number of rides does it become more expensive to go to Tony's Tidal Takeover than Splash?

c. For all three water parks, the cost is a function of the number of rides. Compare the functions for all three water parks in terms of their rate of change. Describe the impact it has on the total cost of attending each park.

Module 5:   Examples of Functions from Geometry

This work is derived from Eureka Math ™ and licensed by Great Minds. ©2015 Great Minds. eureka-math.org
G8-M5-TE-B4-1.3.0-07.2015

EUREKA
MATH

3.   For each part below, leave your answers in terms of π.

   a.   Determine the volume for each three-dimensional figure shown below.

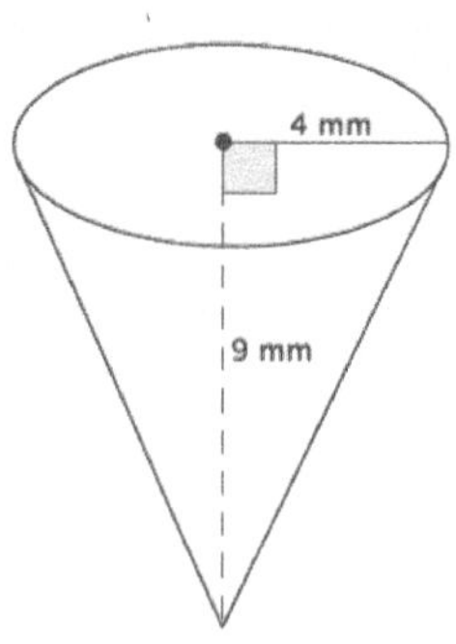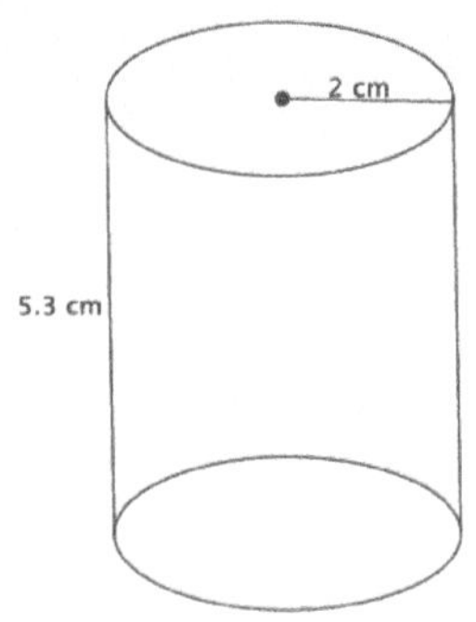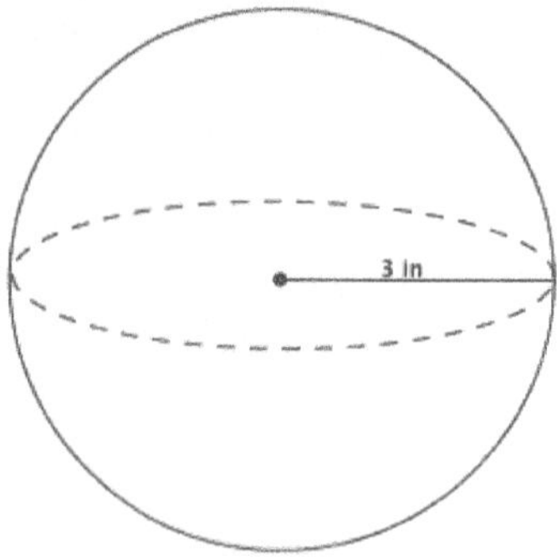

   b.   You want to fill the cylinder shown below with water. All you have is a container shaped like a cone with a radius of 3 inches and a height of 5 inches; you can use this cone-shaped container to take water from a faucet and fill the cylinder. How many cones will it take to fill the cylinder?

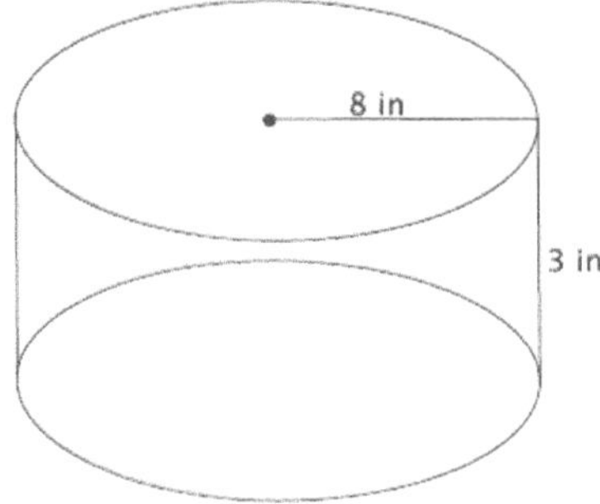

This work is derived from Eureka Math ™ and licensed by Great Minds. ©2015 Great Minds. eureka-math.org
G8-M5-TE-B4-1.3.0-07.2015

c.   You have a cylinder with a diameter of 15 inches and height of 12 inches. What is the volume of the largest sphere that will fit inside of it?

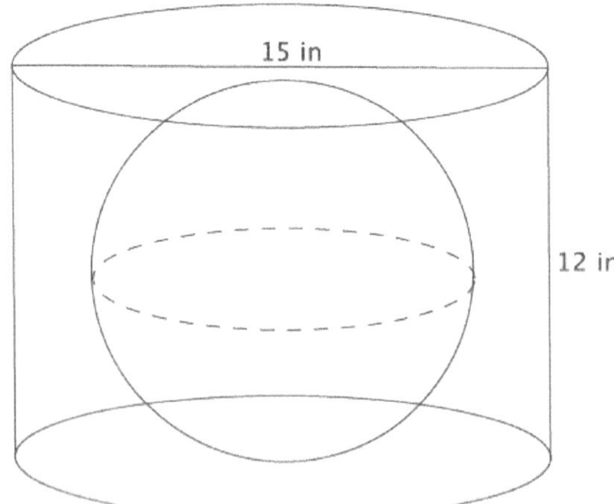

Module 5:   Examples of Functions from Geometry

This work is derived from Eureka Math ™ and licensed by Great Minds. ©2015 Great Minds. eureka-math.org
G8-M5-TE-B4-1.3.0-07.2015

EUREKA
MATH™

A Progression Toward Mastery					
Assessment Task Item		STEP 1 Missing or incorrect answer and little evidence of reasoning or application of mathematics to solve the problem.	STEP 2 Missing or incorrect answer but evidence of some reasoning or application of mathematics to solve the problem.	STEP 3 A correct answer with some evidence of reasoning or application of mathematics to solve the problem, <u>or</u> an incorrect answer with substantial evidence of solid reasoning or application of mathematics to solve the problem.	STEP 4 A correct answer supported by substantial evidence of solid reasoning or application of mathematics to solve the problem.
1	a 8.F.A.1	Student makes little or no attempt to solve the problem.	Student answers at least one of the three questions correctly as 17.3 million, 2011, or yes. Student does not provide an explanation as to why y is a function of x.	Student answers all three questions correctly as 17.3 million, 2011, and yes. Student provides an explanation as to why y is a function of x. Student may not have used vocabulary related to functions.	Student answers all three questions correctly as 17.3 million, 2011, and yes. Student provides a *compelling* explanation as to why y is a function of x and uses appropriate vocabulary related to functions (e.g., assignment, input, and output).
	b 8.F.A.1	Student makes little or no attempt to solve the problem. Student does not write a function or equation. The outputs may or may not be calculated correctly.	Student does not correctly write the equation to describe the function. The outputs may be correct for the function described by the student. The outputs may or may not be calculated correctly. Student may have made calculation errors. Two or more of the outputs are calculated correctly.	Student correctly writes the equation to describe the function as $y = 3x + 4$. Three or more of the outputs are calculated correctly. Student may have made calculation errors.	Student correctly writes the equation to describe the function as $y = 3x + 4$. All four of the outputs are calculated correctly as when $x = -1$, $y = 1$; when $x = \frac{1}{2}$, $y = \frac{11}{2}$; when $x = 1$, $y = 7$; and when $x = 2$, $y = 10$.

Module 5: Examples of Functions from Geometry

This work is derived from Eureka Math ™ and licensed by Great Minds. ©2015 Great Minds. eureka-math.org
G8-M5-TE-B4-1.3.0-07.2015

	c 8.F.A.1	Student makes little or no attempt to solve the problem. Student may have graphed some or all of the input/outputs given.	Student graphs the input/outputs incorrectly (e.g., (4,0) instead of (0,4)). The input/outputs do not appear to be linear.	Student may or may not have graphed the input/outputs correctly (e.g., (4,0) instead of (0,4)). The input/outputs appear to be linear.	Student graphs the input/outputs correctly as (0,4). The input/outputs appear to be linear.
	d 8.F.A.3	Student makes little or no attempt to solve the problem. Student may or may not have made a choice. Student does not give an explanation.	Student incorrectly determines that the meal plan is linear or correctly determines that it is nonlinear. Student does not give an explanation, or the explanation does not include any mathematical reasoning.	Student correctly determines that the meal plan is nonlinear. Explanation includes some mathematical reasoning. Explanation may or may not include reference to the graph.	Student correctly determines that the meal plan is nonlinear. Explanation includes substantial mathematical reasoning. Explanation includes reference to the graph.
2	**a** 8.F.A.2	Student makes little or no attempt to solve the problem. Student may or may not have made a choice. Student does not give an explanation.	Student identifies either choice. Student makes significant calculation errors. Student gives little or no explanation.	Student identifies either choice. Student may have made calculation errors. Explanation may or may not have included the calculation errors.	Student identifies Wally's Water World as the better choice. Student references that for $14 he can ride three rides at Wally's Water World but only two rides at Tony's Tidal Takeover.
	b 8.F.A.2	Student makes little or no attempt to solve the problem. Student does not give an explanation.	Student identifies the number of rides at both parks incorrectly. Student may or may not identify functions to solve the problem. For example, student uses the table or counting method. Student makes some attempt to find the function for one or both of the parks. The functions used are incorrect.	Student identifies the number of rides at one of the parks correctly. Student makes some attempt to identify the function for one or both of the parks. Student may or may not identify functions to solve the problem. For example, student uses the table or counting method. One function used is correct.	Student identifies that the 25th ride at Tony's Tidal Takeover makes it more expensive than Splash. Student may have stated that he could ride 24 rides for $30 at Tony's. Student identifies that the 12th ride at Wally's Water World makes it more expensive than Splash. Student may have stated that he could ride 11 rides for $30 at Wally's. Student identifies functions to solve the problem (e.g., if x is the number of rides, $w = 2x + 8$ for the cost of Wally's, and $t = 0.75x + 12$ for the cost of Tony's).

Module 5: Examples of Functions from Geometry

EUREKA MATH™

This work is derived from Eureka Math ™ and licensed by Great Minds. ©2015 Great Minds. eureka-math.org
G8-M5-TE-B4-1.3.0-07.2015

	c **8.F.A.2**	Student makes little or no attempt to solve the problem.	Student may have identified the rate of change for each park but does so incorrectly. Student may not have compared the rate of change for each park. Student may have described the impact of the rate of change on total cost for one or two of the parks but draws incorrect conclusions.	Student correctly identifies the rate of change for each park. Student may or may not have compared the rate of change for each park. Student may have described the impact of the rate of change on total cost for all parks but makes minor mistakes in the description.	Student correctly identifies the rate of change for each park: Wally's is 2, Tony's is 0.75, and Splash is 0. Student compares the rate of change for each park and identifies which park has the greatest rate of change (or least rate of change) as part of the comparison. Student describes the impact of the rate of change on the total cost for each park.
3	**a** **8.G.C.9**	Student makes little or no attempt to solve the problem. Student finds none or one of the volumes correctly. Student may or may not have included correct units. Student may have omitted π from one or more of the volumes (i.e., the volume of the cone is 48).	Student finds two out of three volumes correctly. Student may or may not have included correct units. Student may have omitted π from one or more of the volumes (i.e., the volume of the cone is 48).	Student finds all three of the volumes correctly. Student does not include the correct units. Student may have omitted π from one or more of the volumes (i.e., the volume of the cone is 48).	Student finds all three of the volumes correctly, that is, the volume of the cone is 48π mm^3, the volume of the cylinder is 21.2π cm^3, and the volume of the sphere is 36π in^3. Student includes the correct units.
	b **8.G.C.9**	Student makes little or no attempt to solve the problem.	Student does not correctly calculate the number of cones. Student makes significant calculation errors. Student may have used the wrong formula for volume of the cylinder or the cone. Student may not have answered in a complete sentence.	Student may have correctly calculated the number of cones, but does not correctly calculate the volume of the cylinder or cone (e.g., volume of the cone is 192, omitting the π). Student correctly calculates the volume of the cone at 15π in^3 or the volume of the cylinder at 192π in^3 but not both. Student may have used incorrect units. Student may have made minor calculation errors. Student may not answer in a complete sentence.	Student correctly calculates that it will take 12.8 cones to fill the cylinder. Student correctly calculates the volume of the cone at 15π in^3 and the volume of the cylinder at 192π in^3. Student answers in a complete sentence.

Module 5:   Examples of Functions from Geometry   **163**

This work is derived from Eureka Math ™ and licensed by Great Minds. ©2015 Great Minds. eureka-math.org
G8-M5-TE-B4-1.3.0-07.2015

c **8.G.C.9**	Student makes little or no attempt to solve the problem.	Student does not correctly calculate the volume. Student may have used the diameter instead of the radius for calculations. Student may have made calculation errors. Student may or may not have omitted π. Student may or may not have included the units.	Student correctly calculates the volume but does not include the units or includes incorrect units (e.g., in^2). Student uses the radius of 6 to calculate the volume. Student may have calculated the volume as 288 (π is omitted).	Student correctly calculates the volume as 288π in^3. Student uses the radius of 6 to calculate the volume. Student includes correct units.

EUREKA
MATH

This work is derived from Eureka Math ™ and licensed by Great Minds. ©2015 Great Minds. eureka-math.org
G8-M5-TE-B4-1.3.0-07.2015

Name ____   Date _______________

1.

a. We define x as a year between 2008 and 2013 and y as the total number of smartphones sold that year, in millions. The table shows values of x and corresponding y values.

Year (x)	2008	2009	2010	2011	2012	2013
Number of smartphones in millions (y)	3.7	17.3	42.4	90	125	153.2

How many smartphones were sold in 2009?

17.3 MILLION SMARTPHONES WERE SOLD IN 2009.

In which year were 90 million smartphones sold?

90 MILLION SMARTPHONES WERE SOLD IN 2011.

Is y a function of x? Explain why or why not.

YES IT IS A FUNCTION BECAUSE FOR EACH INPUT THERE IS EXACTLY ONE OUTPUT. SPECIFICALLY, ONLY ONE NUMBER WILL BE ASSIGNED TO REPRESENT THE NUMBER OF SMARTPHONES SOLD IN THE GIVEN YEAR.

b. Randy began completing the table below to represent a particular linear function. Write an equation to represent the function he was using and compete the table for him.

Input (x)	-3	-1	0	$\frac{1}{2}$	1	2	3
Output (y)	-5	1	4	$\frac{11}{2}$	7	10	13

$y = 3x + 4$

This work is derived from Eureka Math ™ and licensed by Great Minds. ©2015 Great Minds. eureka-math.org
G8-M5-TE-B4-1.3.0-07.2015

c.   Create the graph of the function in part (b).

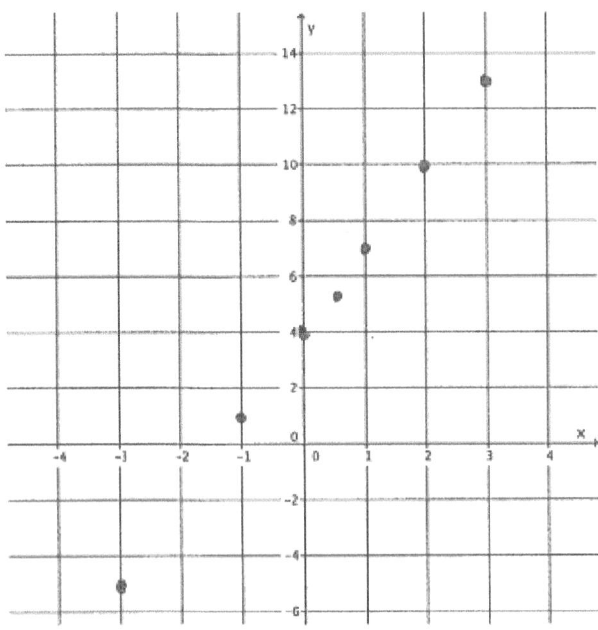

d.   At NYU in 2013, the cost of the weekly meal plan options could be described as a function of the number of meals. Is the cost of the meal plan a linear or nonlinear function? Explain.

8 meals: $125/week
10 meals: $135/week
12 meals: $155/week
21 meals: $220/week

$$\frac{125}{8} = 15.625 \qquad \frac{135}{10} = 13.5 \qquad \frac{155}{12} = 12.917 \qquad \frac{220}{21} = 10.476$$

THE COST OF THE MEAL PLAN IS A NONLINEAR FUNCTION. THE COST PER MEAL IS DIFFERENT BASED ON THE PLAN CHOSEN. FOR EXAMPLE, ONE PLAN CHARGES ALMOST $16 PER MEAL WHILE ANOTHER IS ABOUT $10. ALSO, WHEN THE DATA IS GRAPHED, THE POINTS DO NOT FALL ON A LINE.

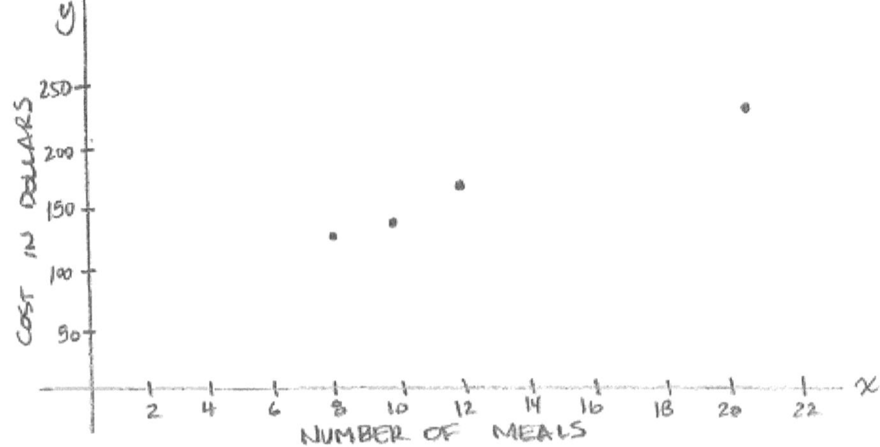

This work is derived from Eureka Math ™ and licensed by Great Minds. ©2015 Great Minds. eureka-math.org
G8-M5-TE-B4-1.3.0-07.2015

2. The cost to enter and go on rides at a local water park, Wally's Water World, is shown in the graph below.

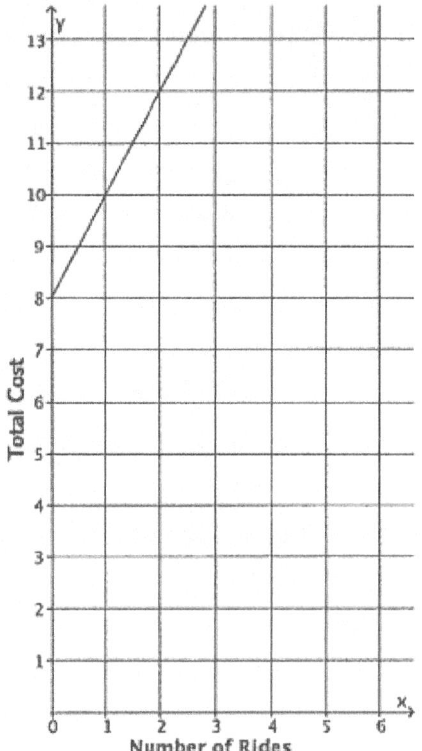

[Handwritten:]
LET x REPRESENT THE NUMBER OF RIDES

LET W REPRESENT THE TOTAL COST AT WALLY'S WATER WORLD

$W = 2x + 8$

A new water park, Tony's Tidal Takeover, just opened. You have not heard anything specific about how much it costs to go to this park, but some of your friends have told you what they spent. The information is organized in the table below.

[Handwritten:] LET x REPRESENT THE NUMBER OF RIDES
LET T REPRESENT THE TOTAL COST AT TONY'S TIDAL TAKEOVER

Number of rides	0	2	4	6
Dollars spent	12.00	13.50	15.00	16.50

[Handwritten between columns:] 1.50   1.50   1.50

Each park charges a different admission fee and a different fee per ride, but the cost of each ride remains the same.

[Handwritten:] $T = 0.75x + 12$

a. If you only have \$14 to spend, which park would you attend (assume the rides are the same quality)? Explain.

[Handwritten:]

WALLY'S
$W = 2x + 8$
$14 = 2x + 8$
$6 = 2x$
$3 = x$

TONY'S
$T = 0.75x + 12$
$14 = 0.75x + 12$
$2 = 0.75x$
$2.67 \approx x$

AT WALLY'S YOU CAN GO ON 3 RIDES WITH \$14, AT TONY'S JUST 2 RIDES. I WOULD GO TO WALLY'S BECAUSE I COULD GO ON MORE RIDES.

This work is derived from Eureka Math ™ and licensed by Great Minds. ©2015 Great Minds. eureka-math.org
G8-M5-TE-B4-1.3.0-07.2015

b. Another water park, Splash, opens and they charge an admission fee of \$30 with no additional fee for rides. At what number of rides does it become more expensive to go to Wally's Water Park than Splash? At what number of rides does it become more expensive to go to Tony's Tidal Takeover than Splash?

LET S REPRESENT TOTAL COST AT SPLASH, $S = 30$.

WALLY'S
$$30 = 2x + 8$$
$$22 = 2x$$
$$11 = x$$

TONY'S
$$30 = 0.75x + 12$$
$$18 = 0.75x$$
$$24 = x$$

AT WALLY'S YOU CAN GO ON 11 RIDES WITH \$30. THE 12TH RIDE MAKES WALLY'S MORE EXPENSIVE THAN SPLASH.

AT TONY'S YOU CAN GO ON 24 RIDES WITH \$30. THE 25TH RIDE MAKES TONY'S MORE EXPENSIVE THAN SPLASH.

c. For all three water parks, the cost is a function of the number of rides. Compare the functions for all three water parks in terms of their rate of change. Describe the impact it has on the total cost of attending each park.

WALLY'S RATE OF CHANGE IS 2, \$2 PER RIDE.
TONY'S RATE OF CHANGE IS 0.75, \$0.75 PER RIDE.
SPLASH'S RATE OF CHANGE IS 0, \$0 EXTRA PER RIDE.
WALLY'S HAS THE GREATEST RATE OF CHANGE. THAT MEANS THAT THE TOTAL COST AT WALLY'S WILL INCREASE THE FASTEST AS WE GO ON MORE RIDES. AT TONY'S, THE RATE OF CHANGE IS JUST 0.75 SO THE TOTAL COST INCREASES WITH THE NUMBER OF RIDES WE GO ON, BUT NOT AS QUICKLY AS WALLY'S. SPLASH HAS A RATE OF CHANGE OF ZERO. THE NUMBER OF RIDES WE GO ON DOES NOT IMPACT THE TOTAL COST AT ALL.

This work is derived from Eureka Math ™ and licensed by Great Minds. ©2015 Great Minds. eureka-math.org
G8-M5-TE-B4-1.3.0-07.2015

EUREKA MATH™

3. For each part below, leave your answers in terms of π.

   a. Determine the volume for each of the three-dimensional figures shown below.

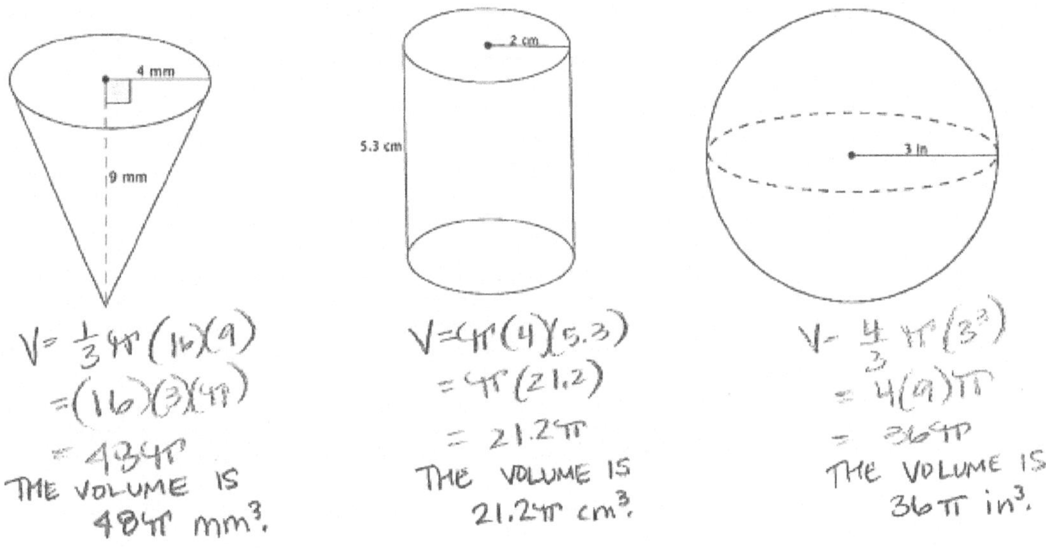

   b. You want to fill the cylinder shown below with water. All you have is a container shaped like a cone with a radius of 3 inches and a height of 5 inches; you can use this cone-shaped container to take water from a faucet and fill the cylinder. How many cones will it take to fill the cylinder?

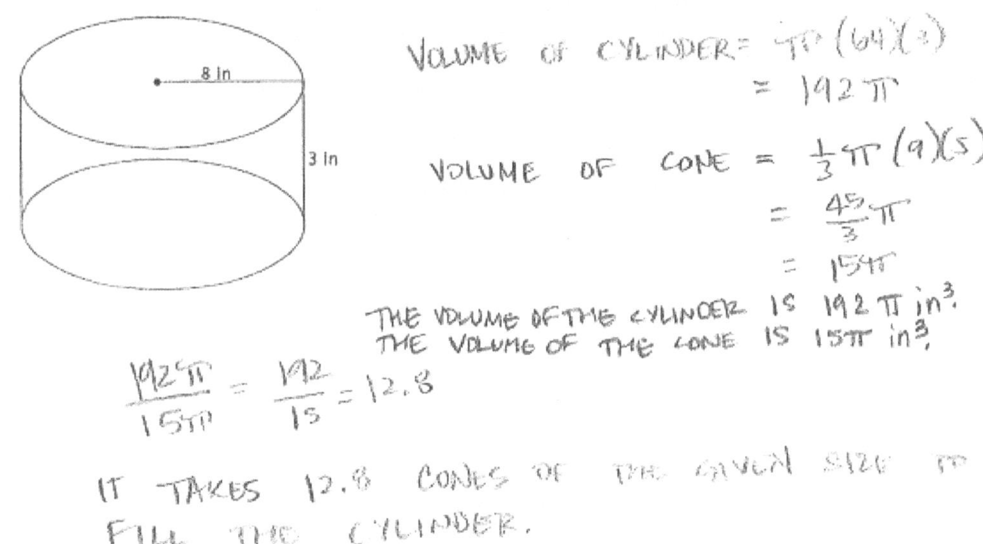

This work is derived from Eureka Math ™ and licensed by Great Minds. ©2015 Great Minds. eureka-math.org
G8-M5-TE-B4-1.3.0-07.2015

c. You have a cylinder with a diameter of 15 cm and height of 12 cm. What is the volume of the largest sphere that will fit inside of it?

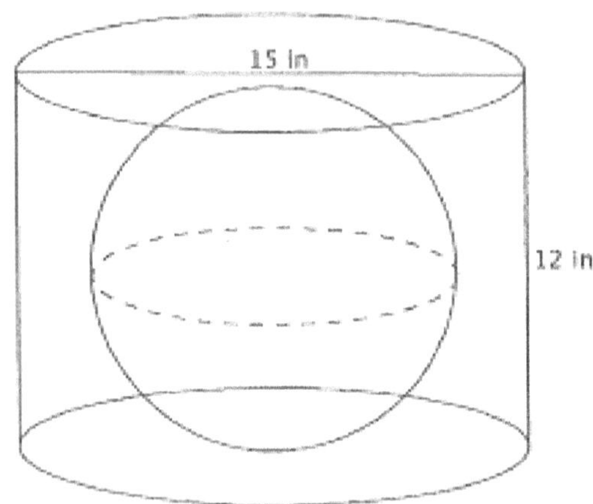

THE CYLINDER HAS RADIUS OF 7.5 cm, BUT THE HEIGHT IS JUST 12cm. THAT MEANS THE MAXIMUM RADIUS FOR THE SPHERE IS 6cm. ANYTHING LARGER WOULD NOT FIT IN THE CYLINDER. THEN THE VOLUME OF THE LARGEST SPHERE THAT WILL FIT IN THE CYLINDER IS 288 in³.

$$V = \frac{4}{3}\pi(6^3)$$
$$= \frac{4}{3}\pi(216)$$
$$= 288\pi$$

EUREKA MATH™

This work is derived from Eureka Math ™ and licensed by Great Minds. ©2015 Great Minds. eureka-math.org
G8-M5-TE-B4-1.3.0-07.2015